2002 Yearbook of Astronomy

2002 Yearbook of Astronomy

edited by

Patrick Moore

associate editor

John Mason

MACMILLAN

First published 2001 by Macmillan
an imprint of Pan Macmillan Ltd
Pan Macmillan, 20 New Wharf Road, London N1 9RR
Basingstoke and Oxford
Associated companies throughout the world
www.panmacmillan.com

ISBN 0 333 90638 1

9 8 7 6 5 4 3 2 1

A CIP catalogue record for this book is available from the British Library.

Typeset by Rowland Phototypesetting Ltd,
Bury St Edmunds, Suffolk
Printed and bound in Great Britain by
Mackays of Chatham plc, Chatham, Kent

Contents

Editor's Foreword

The *2002 Yearbook* follows the usual pattern, but with one important change; this year we have completely new star charts for both the Northern and Southern Hemispheres, drawn by Wil Tirion. As always, Gordon Taylor has provided the material for the monthly notes, and John Isles and Bob Argyle have provided the information on variable stars and double stars, respectively. We also have contributions from our regular authors, such as Paul Murdin, Iain Nicolson and Chris Kitchin, as well as some welcome newcomers. As usual, we have done our best to give you a wide range, both of subject and of technical level. I am also delighted this year to welcome John Mason as my Associate Editor.

PATRICK MOORE
Selsey, July 2001

Preface

New readers will find that all the information in this *Yearbook* is given in diagrammatic or descriptive form; the positions of the planets may easily be found from the specially designed star charts, while the monthly notes describe the movements of the planets and give details of other astronomical phenomena visible in both the northern and southern hemispheres. Two sets of star charts are provided. The **Northern Charts** (pp. 17 to 41) are designed for use at latitude 52°N, but may be used without alteration throughout the British Isles, and (except in the case of eclipses and occultations) in other countries of similar northerly latitude. The **Southern Charts** (pp. 43 to 67) are drawn for latitude 35°S, and are suitable for use in South Africa, Australia and New Zealand, and other locations in approximately the same southerly latitude. The reader who needs more detailed information will find *Norton's Star Atlas* an invaluable guide, while more precise positions of the planets and their satellites, together with predictions of occultations, meteor showers and periodic comets, may be found in the *Handbook* of the British Astronomical Association. Readers will also find details of forthcoming events given in the American monthly magazine *Sky & Telescope*, the British periodical *Astronomy Now* and the Irish magazine *Astronomy & Space*.

Important note

The times given on the star charts and in the Monthly Notes are generally given as local times, using the 24-hour clock, the day beginning at midnight. All the dates, and the times of a few events (e.g. eclipses) are given in Greenwich Mean Time (GMT), which is related to local time by the formula

Local Mean Time = GMT – west longitude

In practice, small differences in longitude are ignored, and the observer will use local clock time, which will be the appropriate Standard (or Zone) Time. As the formula indicates, places in west longitude will have a Standard Time slow on GMT, while places in east longitude will have a Standard Time fast on GMT. As examples we have:

Standard Time in	
New Zealand	GMT + 12 hours
Victoria, NSW	GMT + 10 hours
Western Australia	GMT + 8 hours
South Africa	GMT + 2 hours
British Isles	GMT
Eastern ST	GMT – 5 hours
Central ST	GMT – 6 hours, etc.

If Summer Time is in use, the clocks will have been advanced by one hour, and this hour must be subtracted from the clock time to give Standard Time.

Part I

Monthly Charts and Astronomical Phenomena

Notes on the Star Charts

The stars, together with the Sun, Moon and planets, seem to be set on the surface of the celestial sphere, which appears to rotate about the Earth from east to west. Since it is impossible to represent a curved surface accurately on a plane, any kind of star map is bound to contain some form of distortion.

Most of the monthly star charts which appear in the various journals and some national newspapers are drawn in circular form. This is perfectly accurate, but it can make the charts awkward to use. For the star charts in this volume, we have preferred to give two hemispherical maps for each month of the year, one showing the northern aspect of the sky and the other showing the southern aspect. Two sets of monthly charts are provided; one for observers in the Northern Hemisphere and one for those in the Southern Hemisphere.

Unfortunately, the constellations near the overhead point (the zenith) on these hemispherical charts can be rather distorted. This would be a serious drawback for precision charts, but what we have done is to give maps which are best suited to star recognition. We have also refrained from putting in too many stars, so that the main patterns stand out clearly. To help observers with any distortions near the zenith, and the lack of overlap between the charts of each pair, we have also included two circular maps, one showing all the constellations in the northern half of the sky, and one those in the southern half. Incidentally, there is a curious illusion that stars at an altitude of 60° or more are actually overhead, and beginners may often feel that they are leaning over backwards in trying to see them.

The charts show all stars down to the fourth magnitude, together with a number of fainter stars which are necessary to define the shapes of constellations. There is no standard system for representing the outlines of the constellations, and triangles and other simple figures have been used to give outlines which are easy to trace with the naked eye. The names of the constellations are given, together with the proper names of the brighter stars. The apparent magnitudes of the stars are indicated roughly by using different sizes of dot, the larger dots representing the brighter stars.

The two sets of star charts – one each for Northern and Southern Hemisphere observers – are similar in design. At each opening there is a single circular chart which shows all the constellations in that hemisphere of the sky. (These two charts are centred on the North and South Celestial Poles, respectively.) Then there are twelve double-page spreads, showing the northern and southern aspects for each month of the year for observers in that hemisphere. In the **Northern Charts** (drawn for latitude 52°N) the left-hand chart of each spread shows the northern half of the sky (lettered 1N, 2N, 3N . . . 12N), and the corresponding right-hand chart shows the southern half of the sky (lettered 1S, 2S, 3S . . . 12S). The arrangement and lettering of the charts is exactly the same for the **Southern Charts** (drawn for latitude 35°S).

Because the sidereal day is shorter than the solar day, the stars appear to rise and set about four minutes earlier each day, and this amounts to two hours in a month. Hence the twelve pairs of charts in each set are sufficient to give the appearance of the sky throughout the day at intervals of two hours, or at the same time of night at monthly intervals throughout the year. For example, charts 1N and 1S here are drawn for 23 hours on January 6. The view will also be the same on October 6 at 05 hours; November 6 at 03 hours; December 6 at 01 hours and February 6 at 21 hours. The actual range of dates and times when the stars on the charts are visible is indicated on each page. Each pair of charts is numbered in bold type, and the number to be used for any given month and time may be found from the following table:

Local Time	**18^h**	**20^h**	**22^h**	**0^h**	**2^h**	**4^h**	**6^h**
January	11	12	1	2	3	4	5
February	12	1	2	3	4	5	6
March	1	2	3	4	5	6	7
April	2	3	4	5	6	7	8
May	3	4	5	6	7	8	9
June	4	5	6	7	8	9	10
July	5	6	7	8	9	10	11
August	6	7	8	9	10	11	12
September	7	8	9	10	11	12	1
October	8	9	10	11	12	1	2

Local Time	**18h**	**20h**	**22h**	**0h**	**2h**	**4h**	**6h**
November	9	10	11	12	1	2	3
December	10	11	12	1	2	3	4

On these charts, the ecliptic is drawn as a broken line on which longitude is marked every 10°. The positions of the planets are then easily found by reference to the table on p. 73. It will be noticed that on the **Southern Charts** the *ecliptic* may reach an altitude in excess of 62½° on the star charts showing the northern aspect (5N to 9N). The continuations of the broken line will be found on the corresponding charts for the southern aspect (5S, 6S, 8S and 9S).

Northern Star Charts

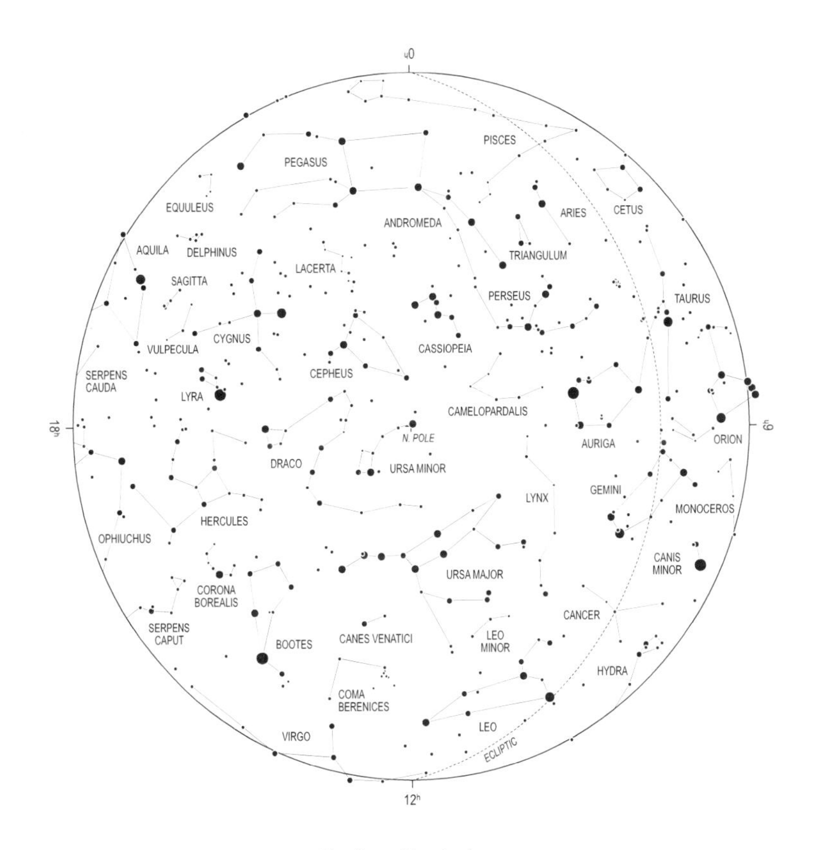

Northern Hemisphere

Note that the markers at 0^h, 6^h, 12^h and 18^h indicate hours of Right Ascension.

1N

October 6 at 5h
November 6 at 3h
December 6 at 1h
January 6 at 23h
February 6 at 21h

October 21 at 4h
November 21 at 2h
December 21 at midnight
January 21 at 22h
February 21 at 20h

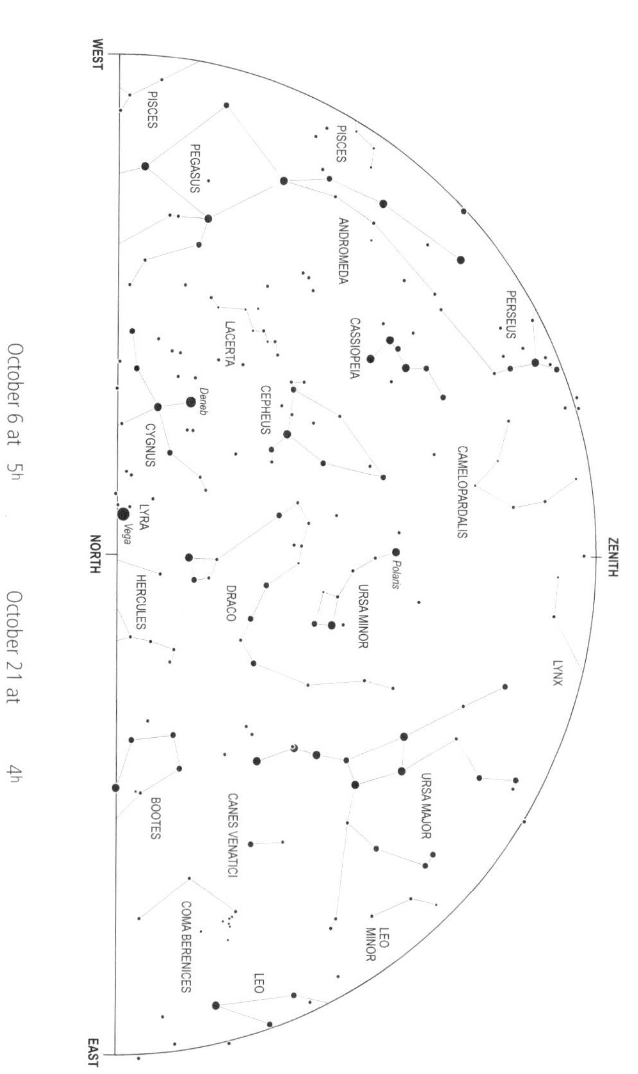

1S

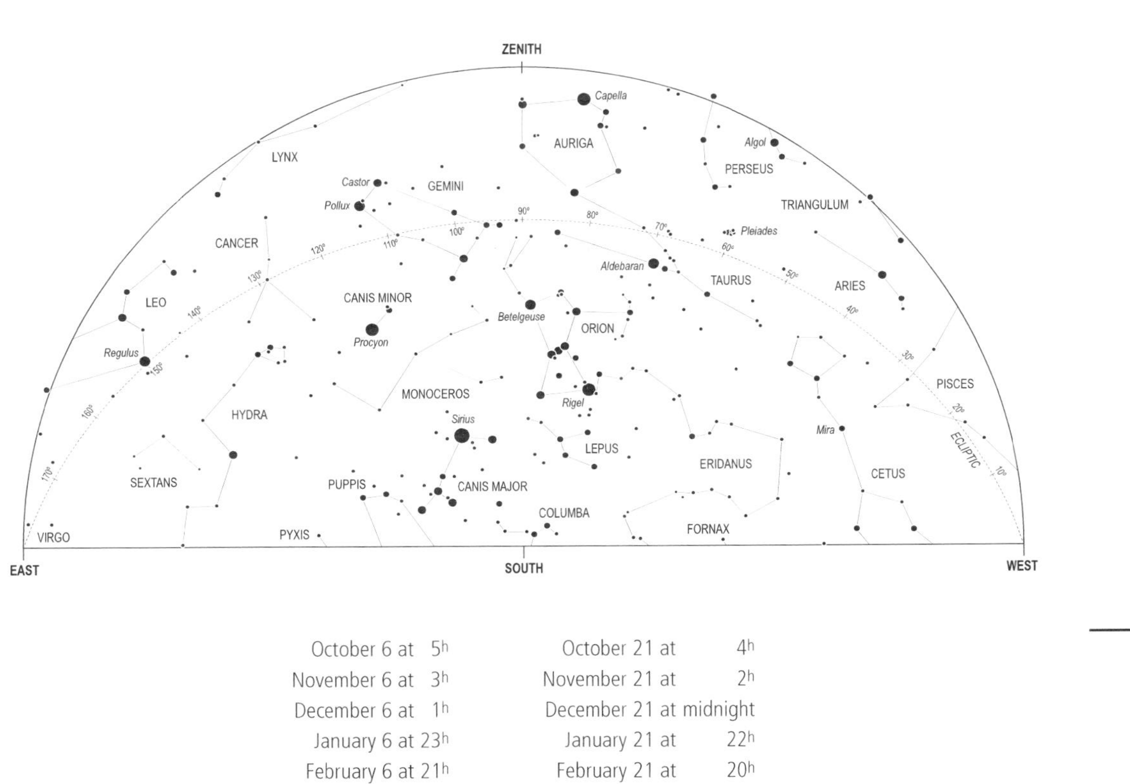

October 6 at 5^h	October 21 at 4^h
November 6 at 3^h	November 21 at 2^h
December 6 at 1^h	December 21 at midnight
January 6 at 23^h	January 21 at 22^h
February 6 at 21^h	February 21 at 20^h

2N

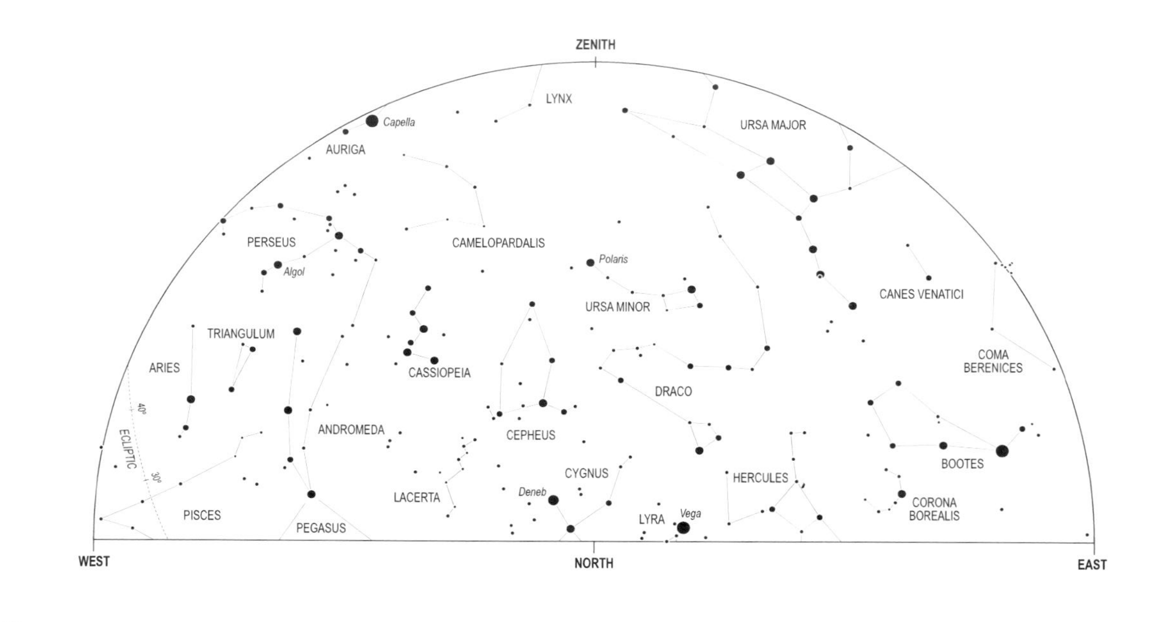

November 6 at 5h	November 21 at 4h
December 6 at 3h	December 21 at 2h
January 6 at 1h	January 21 at midnight
February 6 at 23h	February 21 at 22h
March 6 at 21h	March 21 at 20h

2S

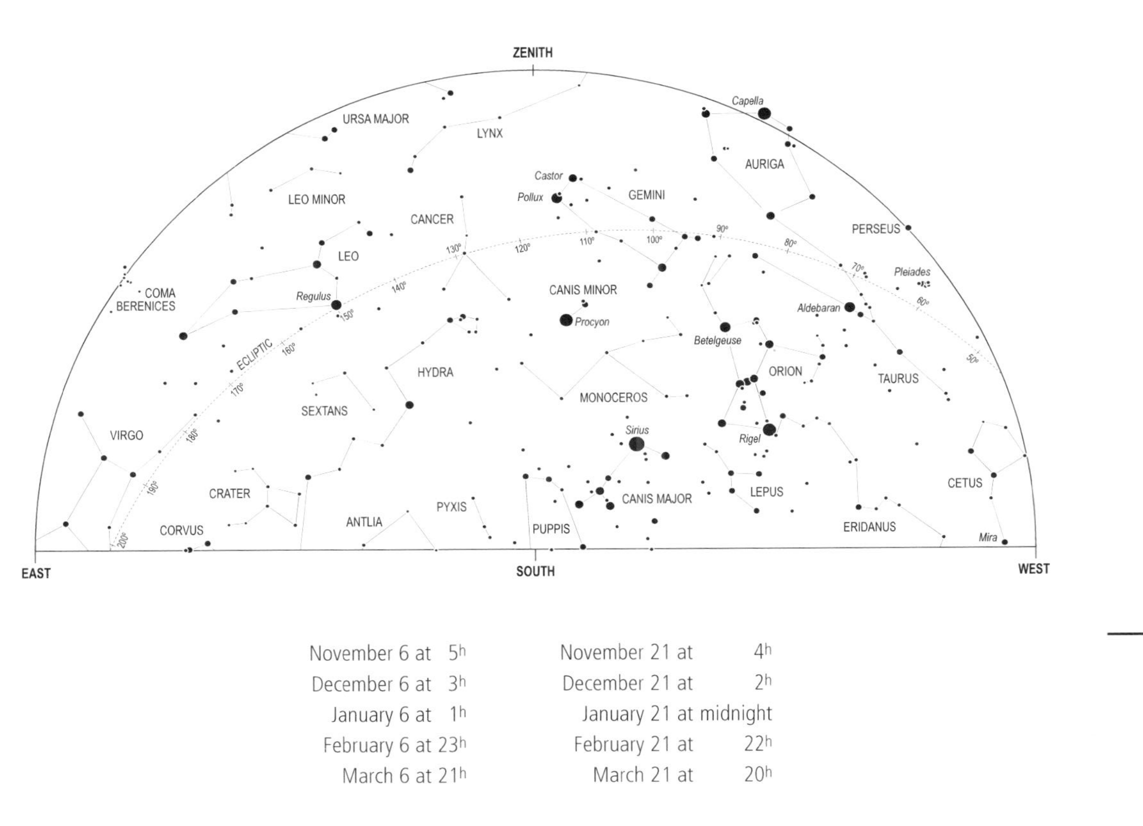

November 6 at 5h	November 21 at 4h
December 6 at 3h	December 21 at 2h
January 6 at 1h	January 21 at midnight
February 6 at 23h	February 21 at 22h
March 6 at 21h	March 21 at 20h

3N

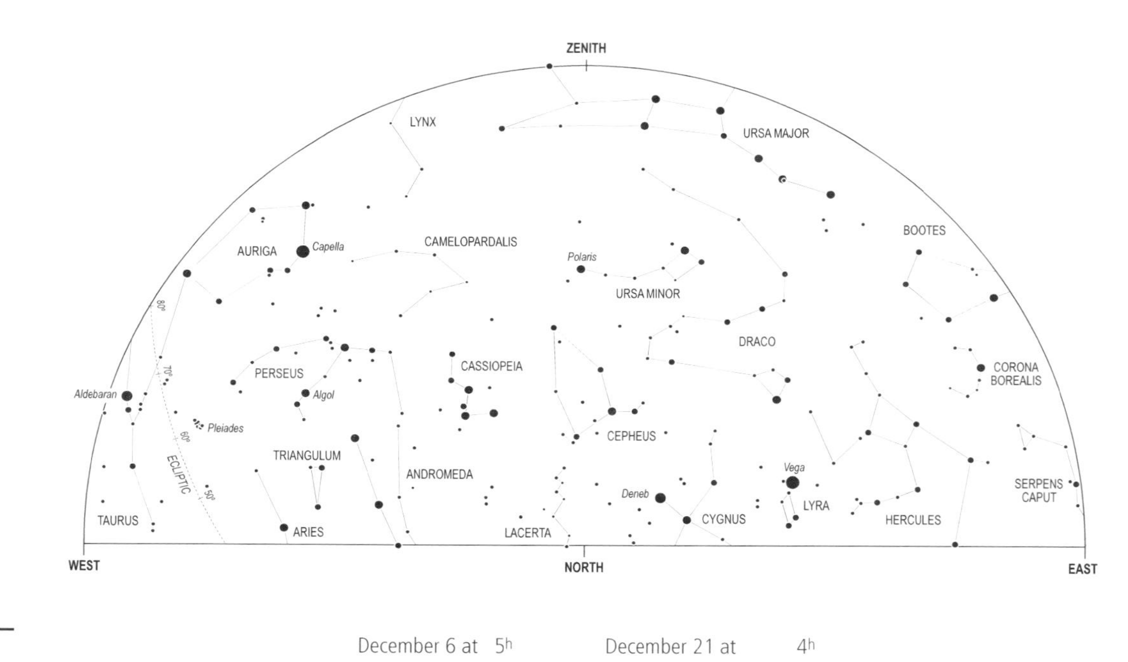

December 6 at 5h	December 21 at 4h
January 6 at 3h	January 21 at 2h
February 6 at 1h	February 21 at midnight
March 6 at 23h	March 21 at 22h
April 6 at 21h	April 21 at 20h

3S

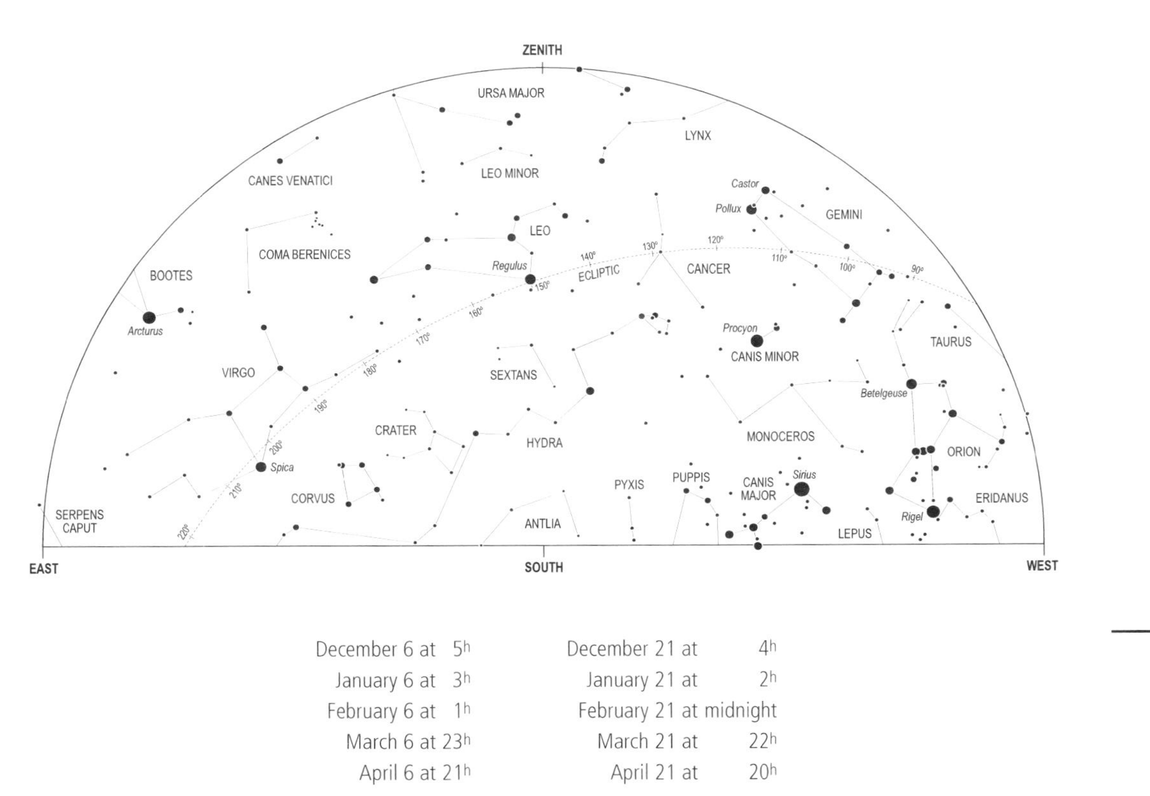

December 6 at 5h	December 21 at 4h
January 6 at 3h	January 21 at 2h
February 6 at 1h	February 21 at midnight
March 6 at 23h	March 21 at 22h
April 6 at 21h	April 21 at 20h

4N

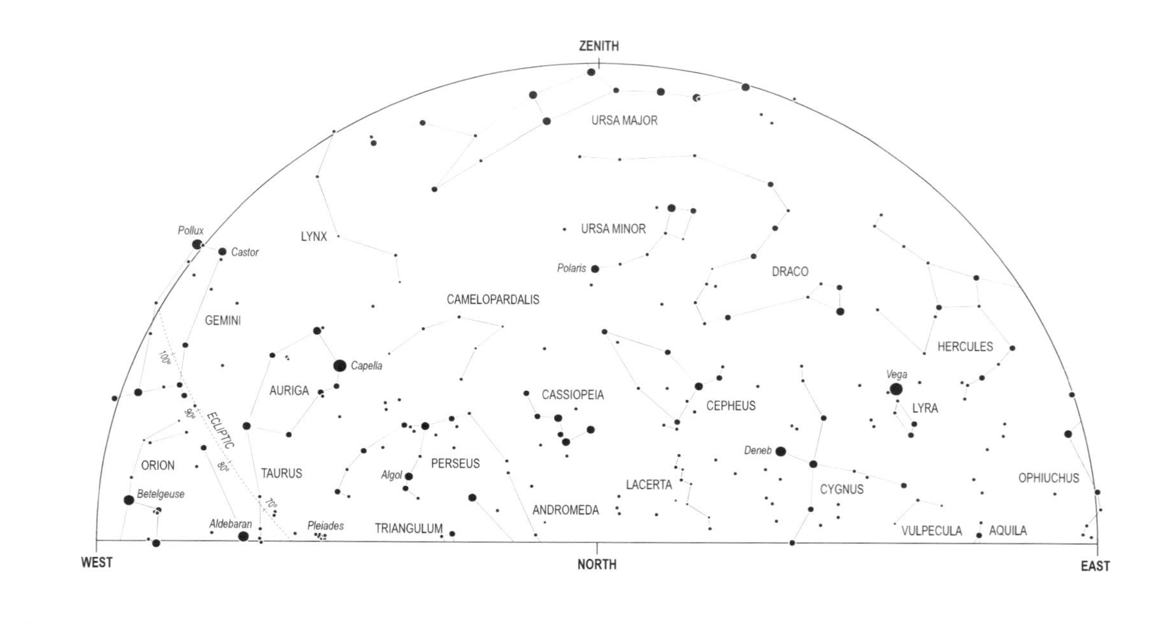

January 6 at 5h	January 21 at 4h
February 6 at 3h	February 21 at 2h
March 6 at 1h	March 21 at midnight
April 6 at 23h	April 21 at 22h
May 6 at 21h	May 21 at 20h

4S

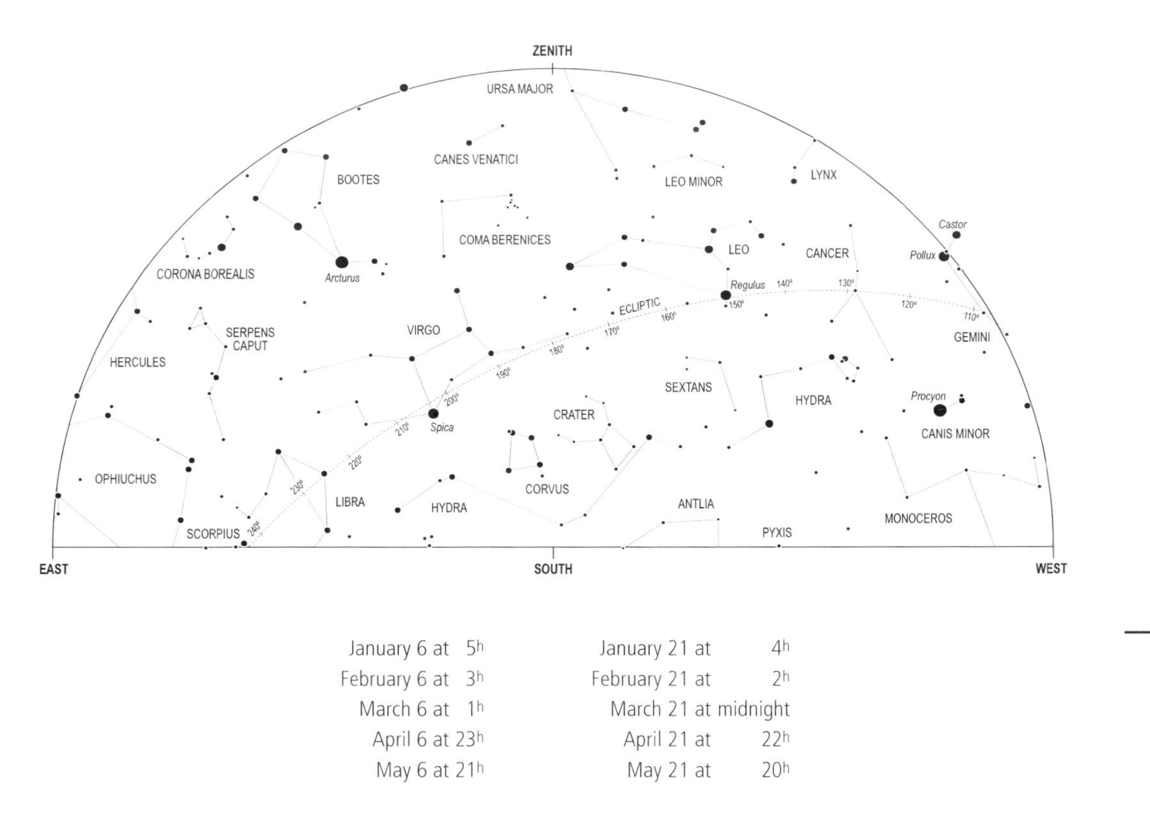

January 6 at 5h	January 21 at 4h
February 6 at 3h	February 21 at 2h
March 6 at 1h	March 21 at midnight
April 6 at 23h	April 21 at 22h
May 6 at 21h	May 21 at 20h

5N

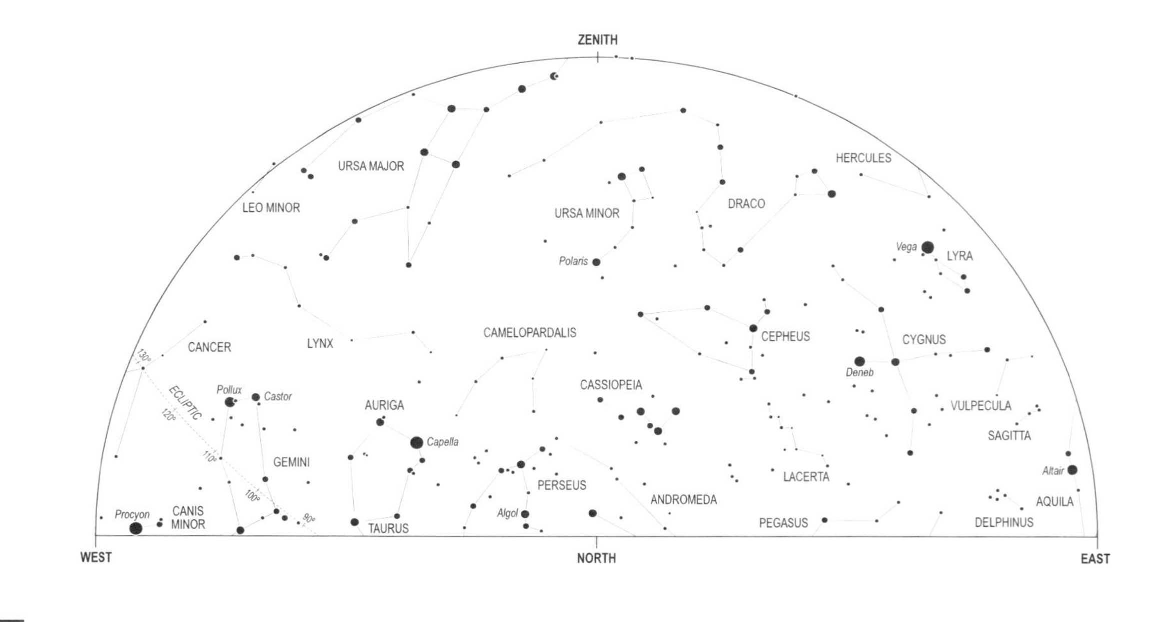

January 6 at 7h	January 21 at 6h
February 6 at 5h	February 21 at 4h
March 6 at 3h	March 21 at 2h
April 6 at 1h	April 21 at midnight
May 6 at 23h	May 21 at 22h

5S

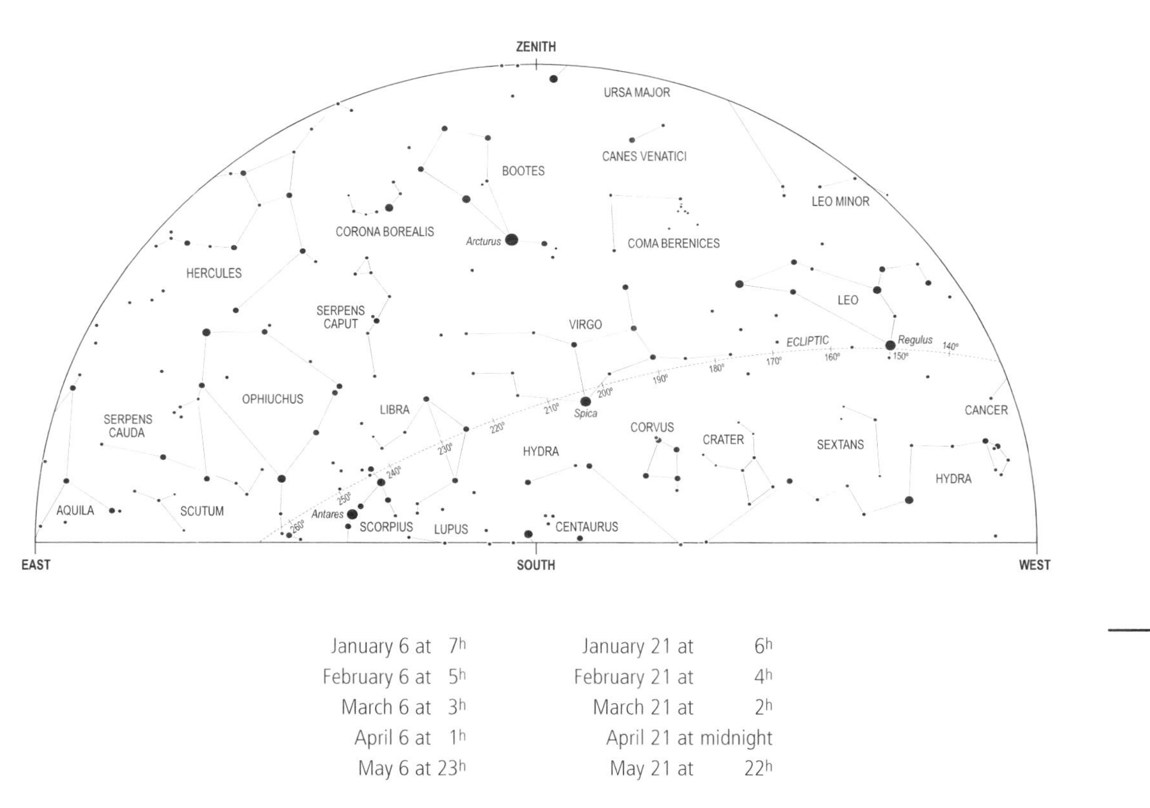

January 6 at 7^{h}	January 21 at 6^{h}
February 6 at 5^{h}	February 21 at 4^{h}
March 6 at 3^{h}	March 21 at 2^{h}
April 6 at 1^{h}	April 21 at midnight
May 6 at 23^{h}	May 21 at 22^{h}

6N

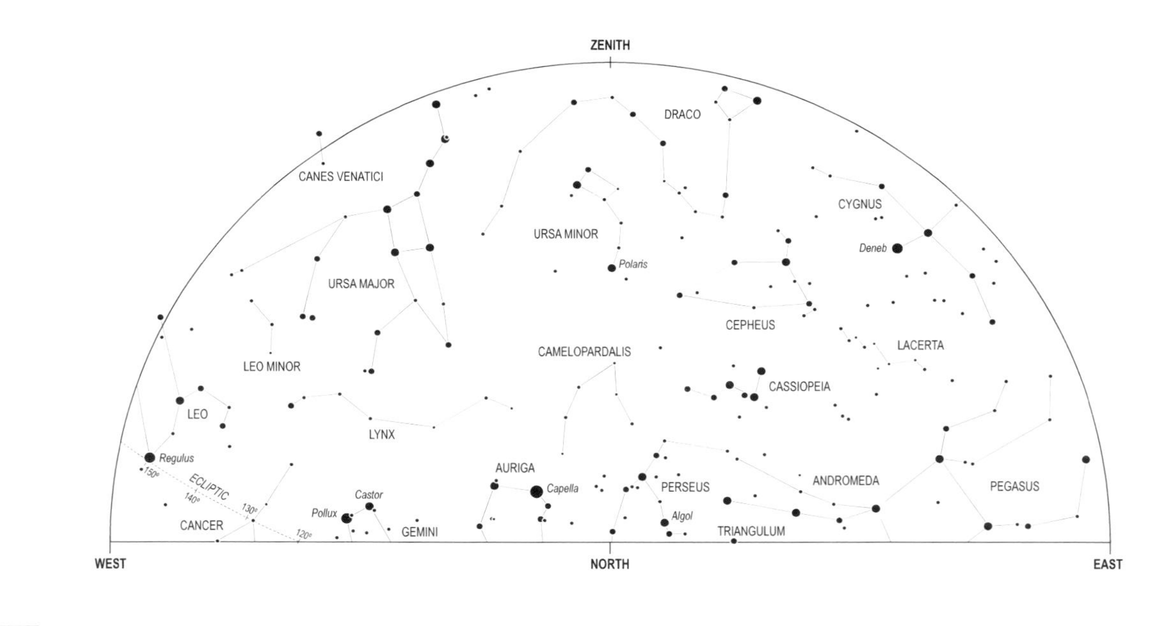

March 6 at 5^h	March 21 at 4^h
April 6 at 3^h	April 21 at 2^h
May 6 at 1^h	May 21 at midnight
June 6 at 23^h	June 21 at 22^h
July 6 at 21^h	July 21 at 20^h

6S

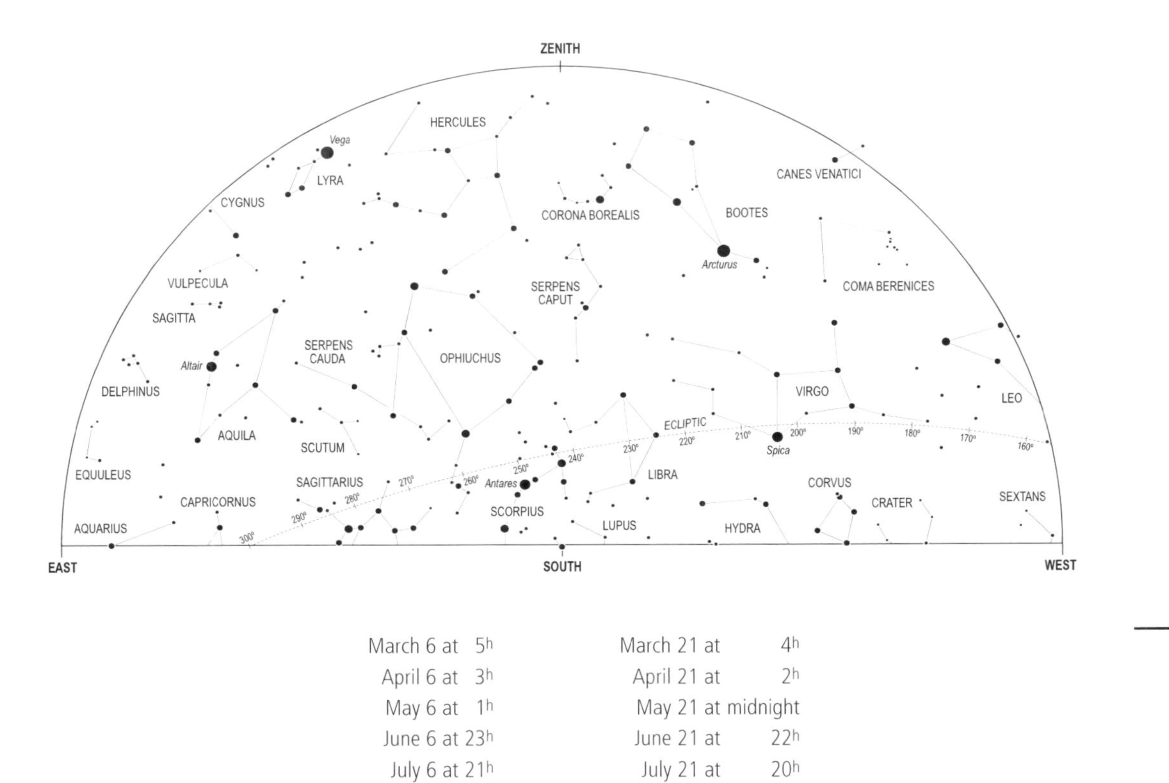

March 6 at 5h	March 21 at 4h
April 6 at 3h	April 21 at 2h
May 6 at 1h	May 21 at midnight
June 6 at 23h	June 21 at 22h
July 6 at 21h	July 21 at 20h

7N

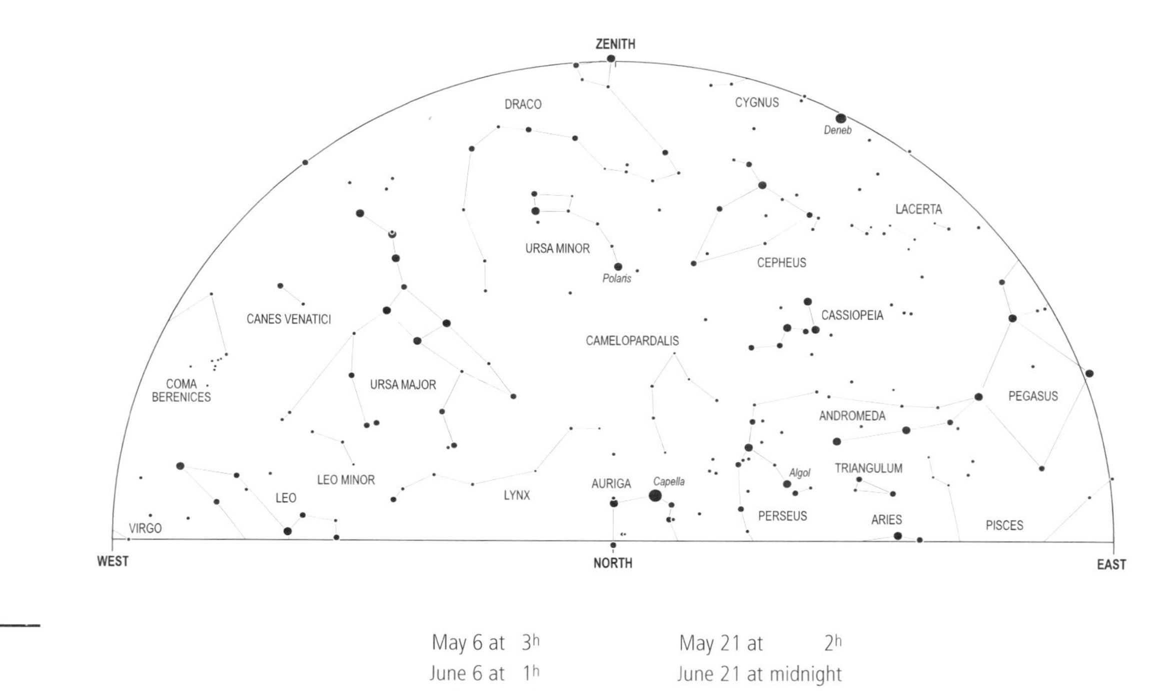

May 6 at 3^h	May 21 at 2^h
June 6 at 1^h	June 21 at midnight
July 6 at 23^h	July 21 at 22^h
August 6 at 21^h	August 21 at 20^h
September 6 at 19^h	September 21 at 18^h

7S

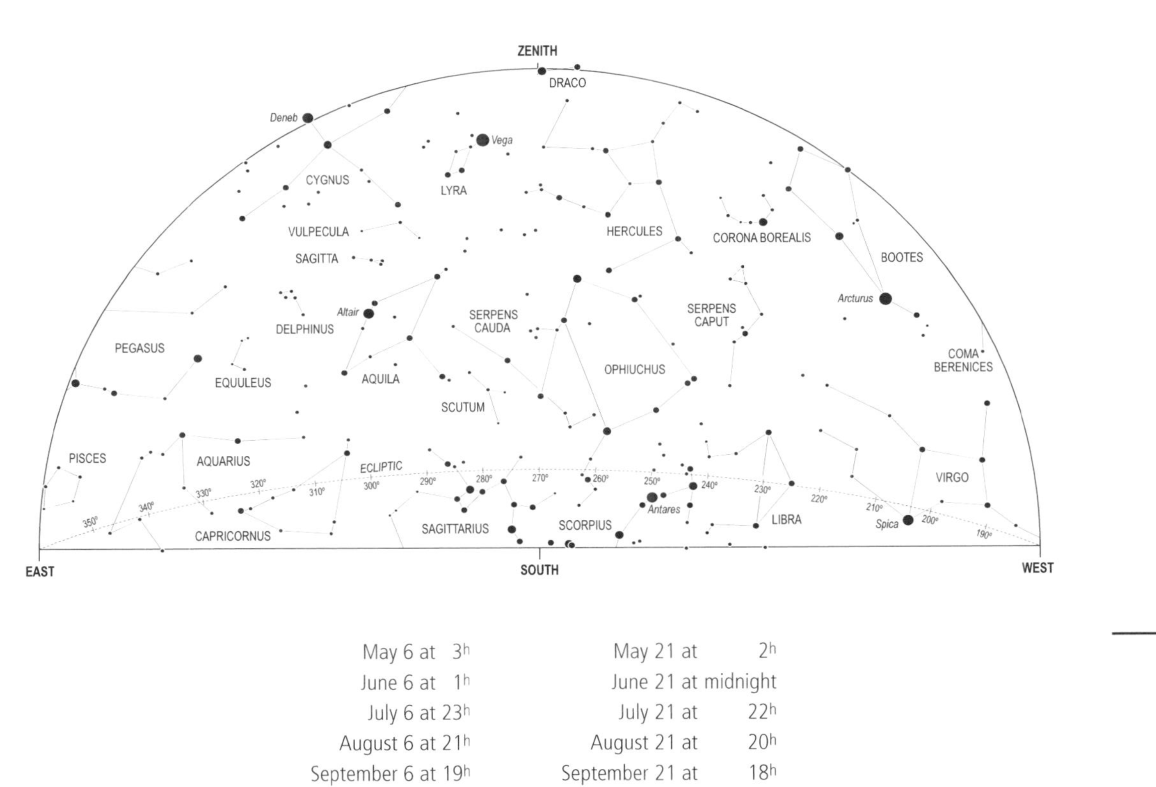

May 6 at 3h	May 21 at 2h
June 6 at 1h	June 21 at midnight
July 6 at 23h	July 21 at 22h
August 6 at 21h	August 21 at 20h
September 6 at 19h	September 21 at 18h

8N

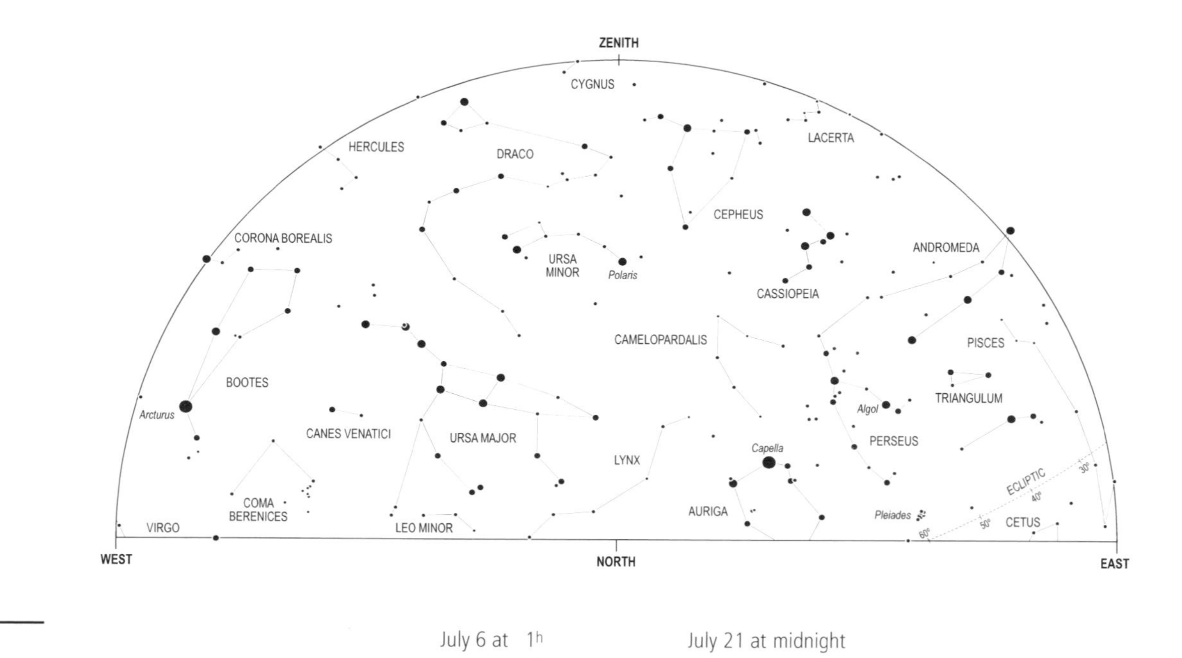

July 6 at 1h	July 21 at midnight
August 6 at 23h	August 21 at 22h
September 6 at 21h	September 21 at 20h
October 6 at 19h	October 21 at 18h
November 6 at 17h	November 21 at 16h

8S

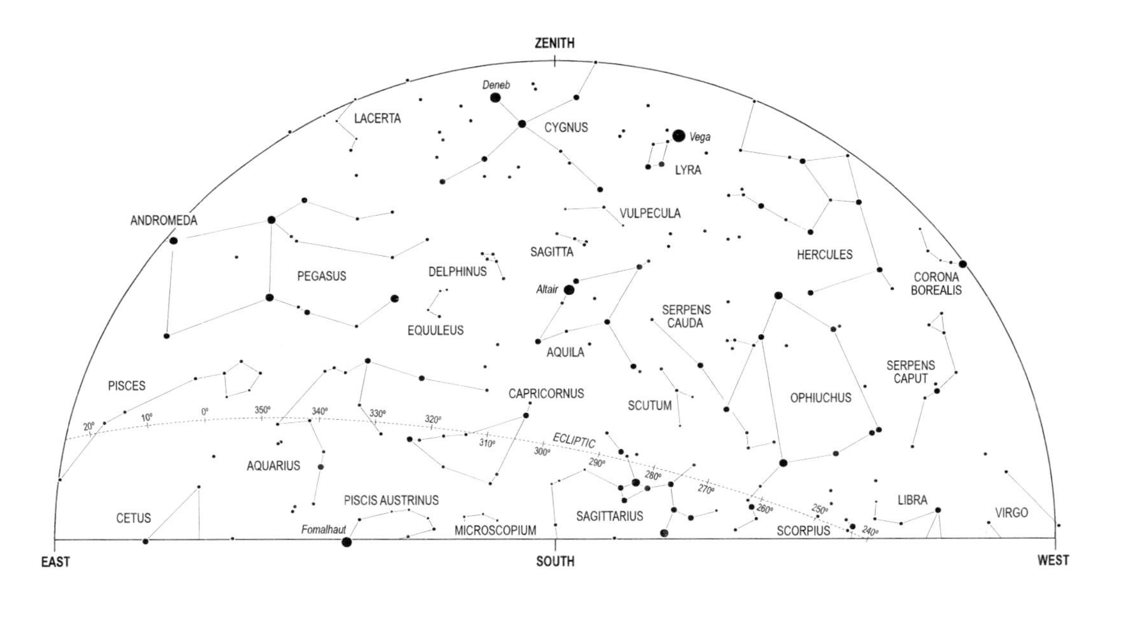

July 6 at 1h	July 21 at midnight
August 6 at 23h	August 21 at 22h
September 6 at 21h	September 21 at 20h
October 6 at 19h	October 21 at 18h
November 6 at 17h	November 21 at 16h

9N

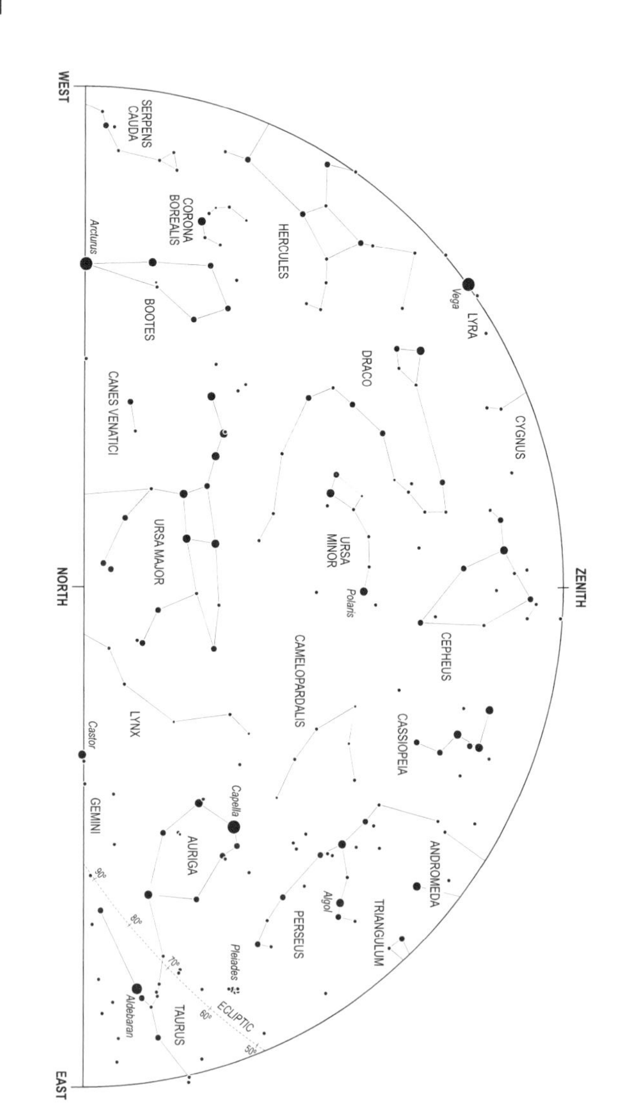

August 6 at 1h	August 21 at midnight
September 6 at 23h	September 21 at 22h
October 6 at 21h	October 21 at 20h
November 6 at 19h	November 21 at 18h
December 6 at 17h	December 21 at 16h

9S

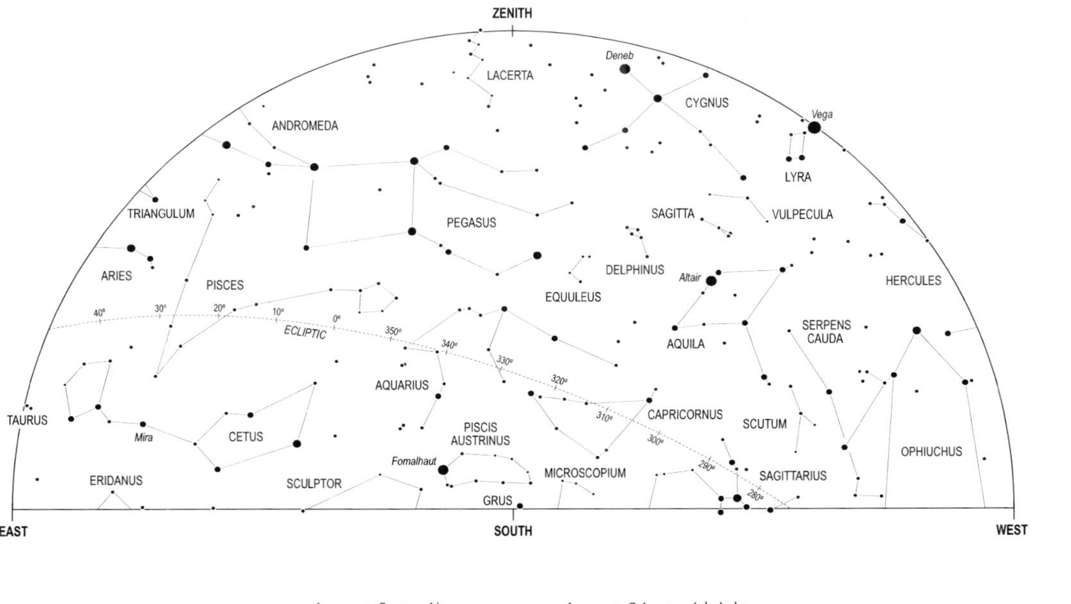

August 6 at 1h	August 21 at midnight
September 6 at 23h	September 21 at 22h
October 6 at 21h	October 21 at 20h
November 6 at 19h	November 21 at 18h
December 6 at 17h	December 21 at 16h

10N

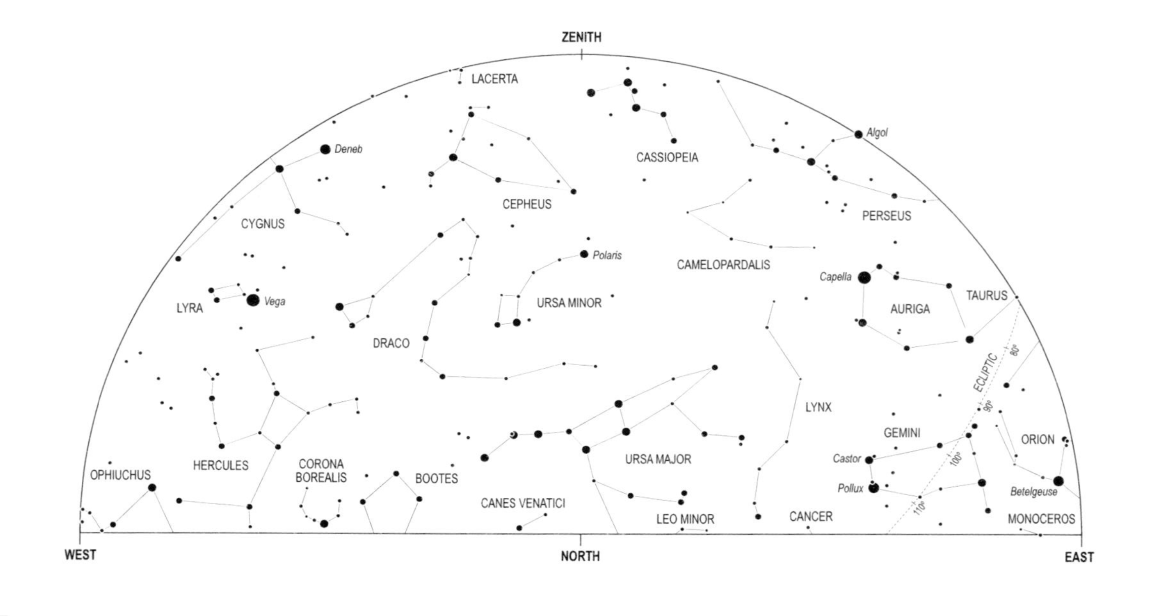

August 6 at 3h	August 21 at 2h
September 6 at 1h	September 21 at midnight
October 6 at 23h	October 21 at 22h
November 6 at 21h	November 21 at 20h
December 6 at 19h	December 21 at 18h

10S

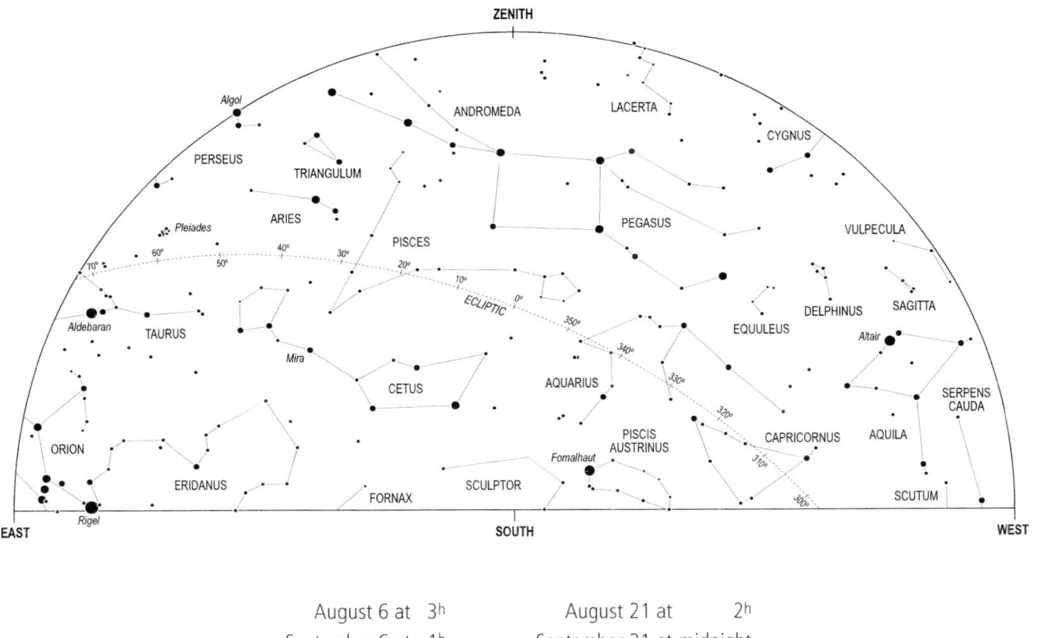

August 6 at 3h	August 21 at 2h
September 6 at 1h	September 21 at midnight
October 6 at 23h	October 21 at 22h
November 6 at 21h	November 21 at 20h
December 6 at 19h	December 21 at 18h

11N

September 6 at 3h | September 21 at 2h
October 6 at 1h | October 21 at midnight
November 6 at 23h | November 21 at 22h
December 6 at 21h | December 21 at 20h
January 6 at 19h | January 21 at 18h

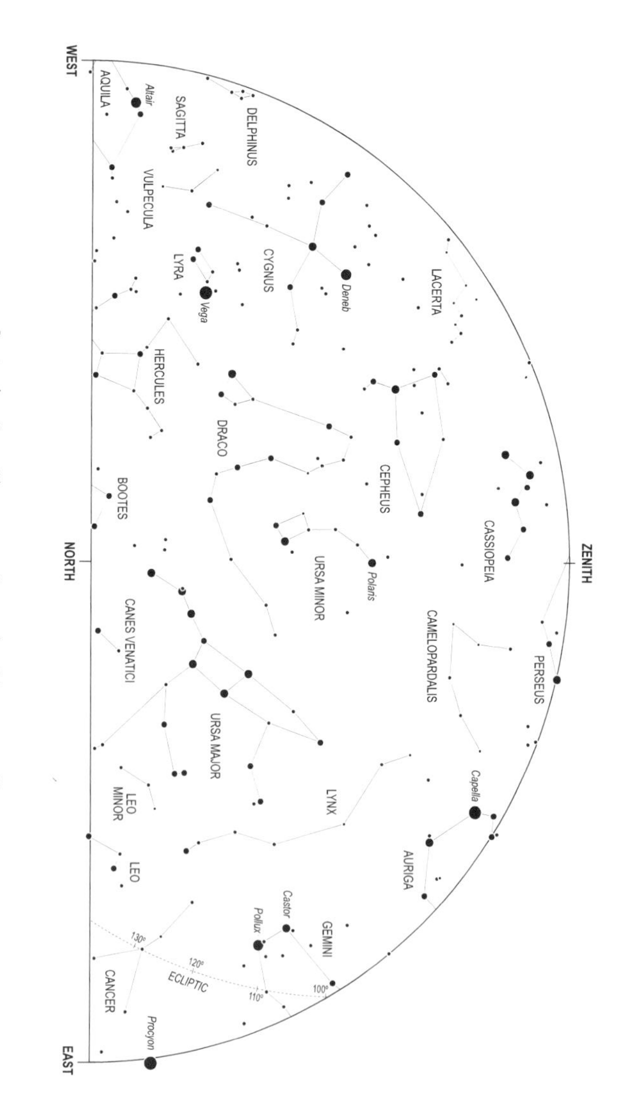

11S

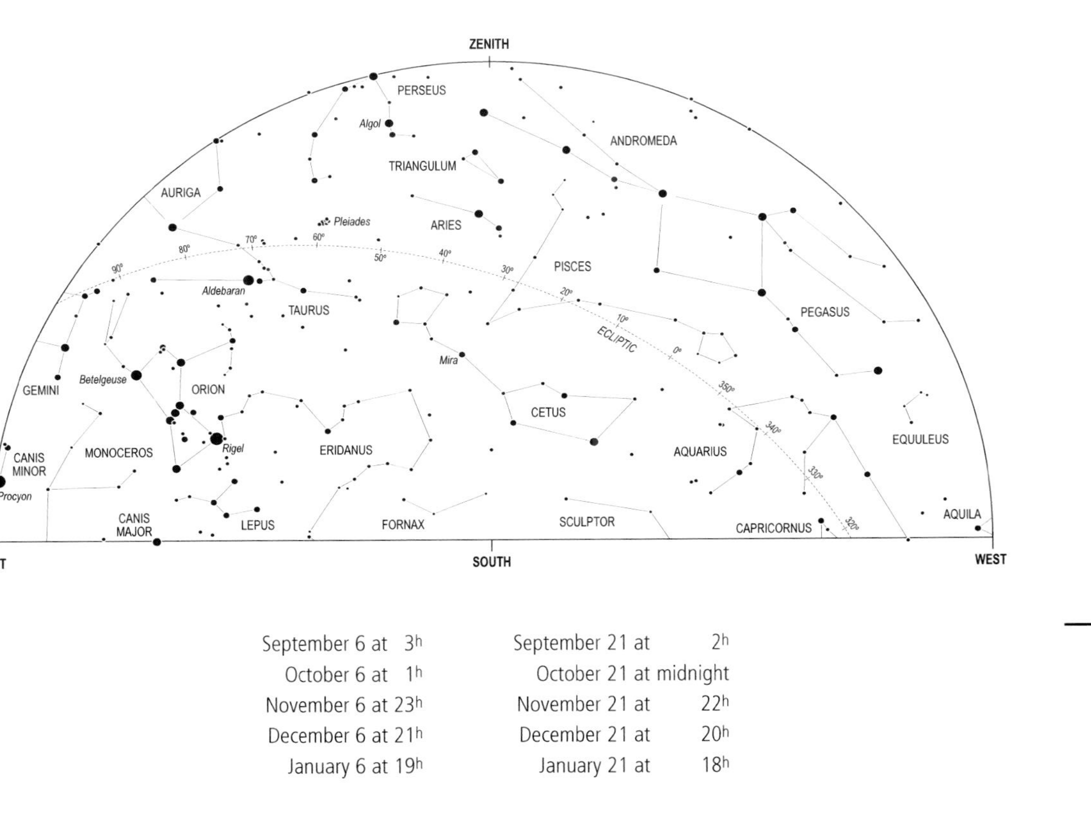

September 6 at 3h	September 21 at 2h
October 6 at 1h	October 21 at midnight
November 6 at 23h	November 21 at 22h
December 6 at 21h	December 21 at 20h
January 6 at 19h	January 21 at 18h

12N

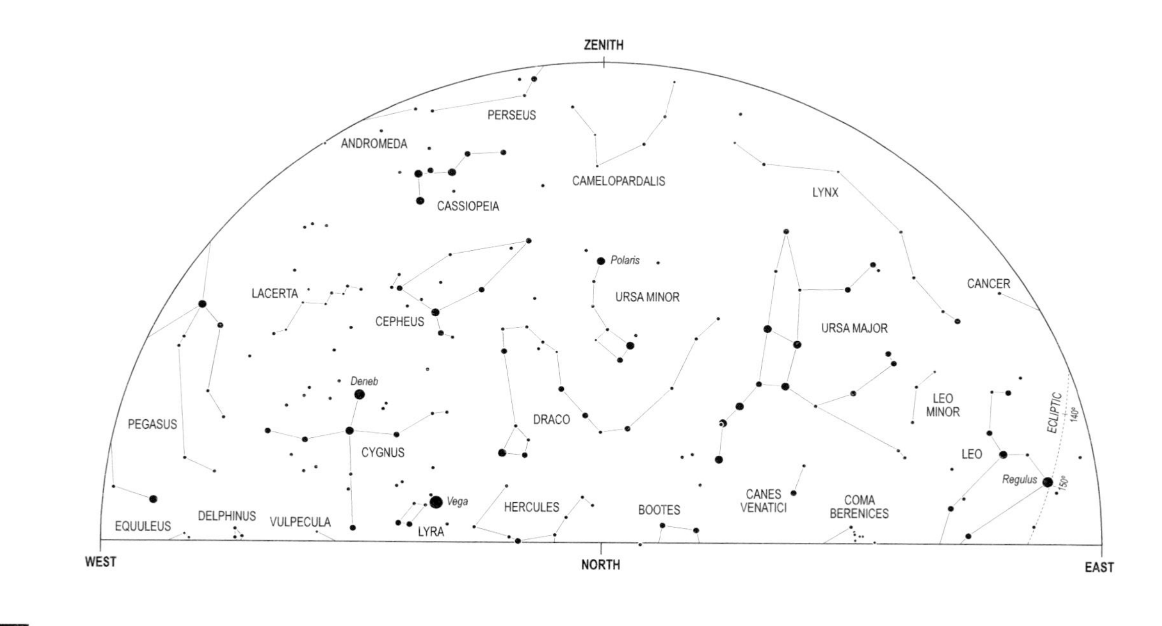

October 6 at 3^h	October 21 at 2^h
November 6 at 1^h	November 21 at midnight
December 6 at 23^h	December 21 at 22^h
January 6 at 21^h	January 21 at 20^h
February 6 at 19^h	February 21 at 18^h

12S

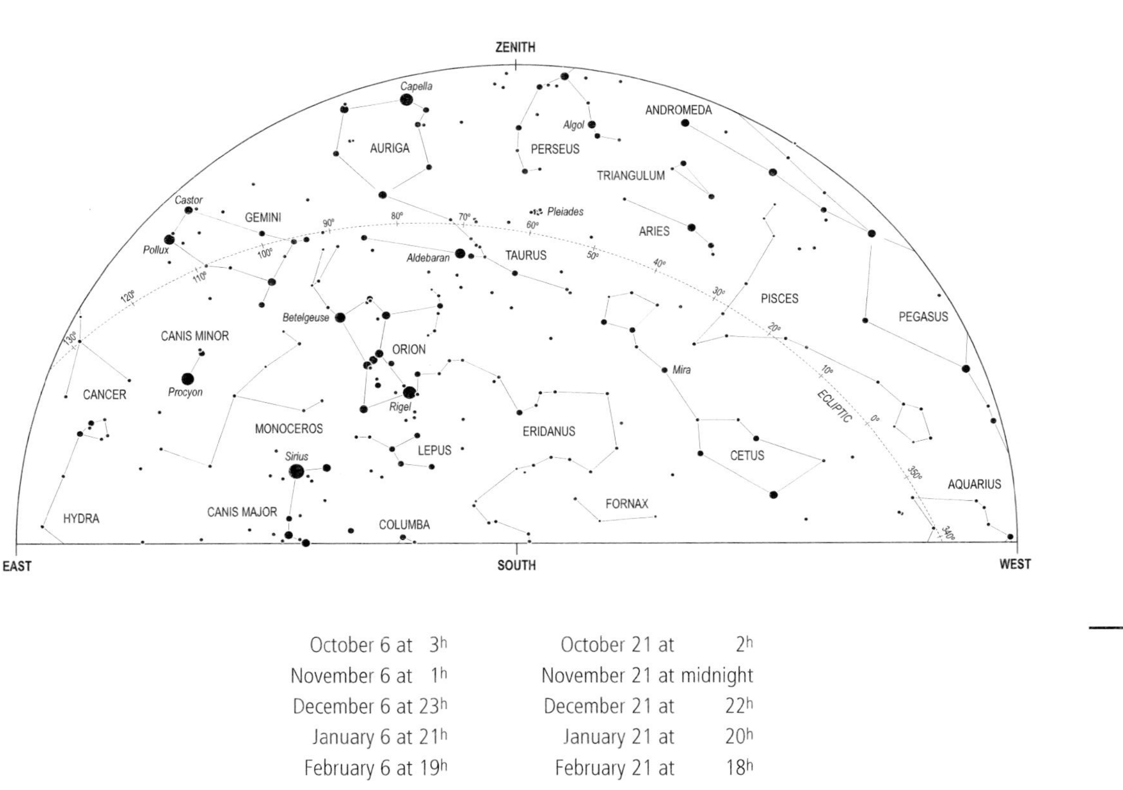

October 6 at 3h	October 21 at 2h
November 6 at 1h	November 21 at midnight
December 6 at 23h	December 21 at 22h
January 6 at 21h	January 21 at 20h
February 6 at 19h	February 21 at 18h

Southern Star Charts

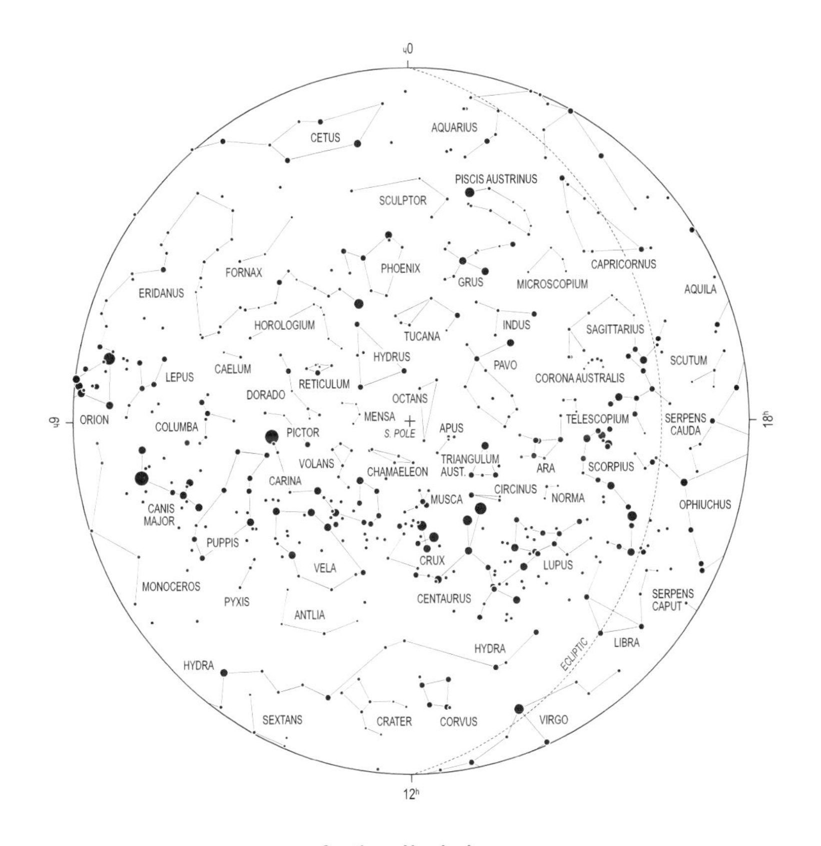

Southern Hemisphere

Note that the markers at 0h, 6h, 12h and 18h indicate hours of Right Ascension.

1N

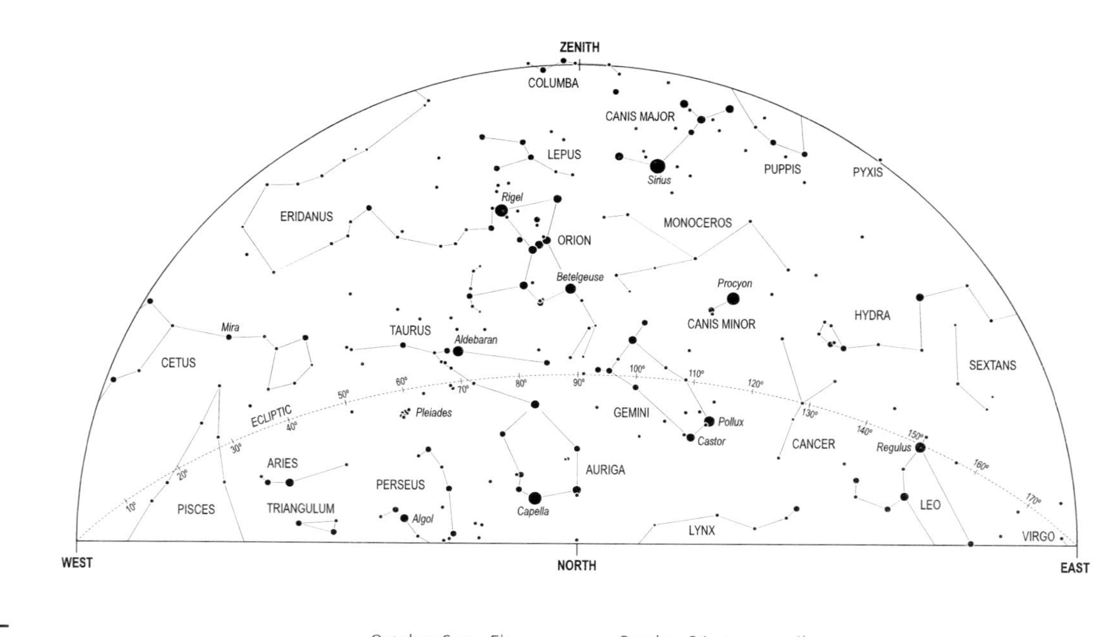

October 6 at 5h	October 21 at 4h
November 6 at 3h	November 21 at 2h
December 6 at 1h	December 21 at midnight
January 6 at 23h	January 21 at 22h
February 6 at 21h	February 21 at 20h

1S

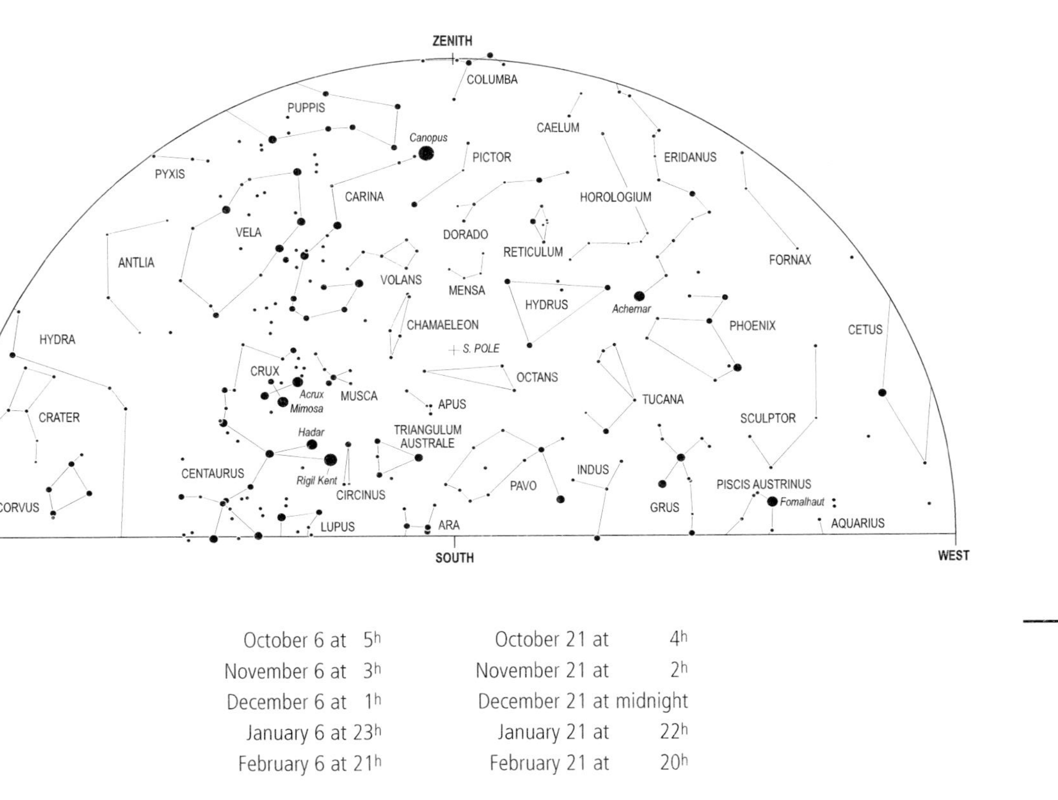

October 6 at 5h	October 21 at 4h
November 6 at 3h	November 21 at 2h
December 6 at 1h	December 21 at midnight
January 6 at 23h	January 21 at 22h
February 6 at 21h	February 21 at 20h

2N

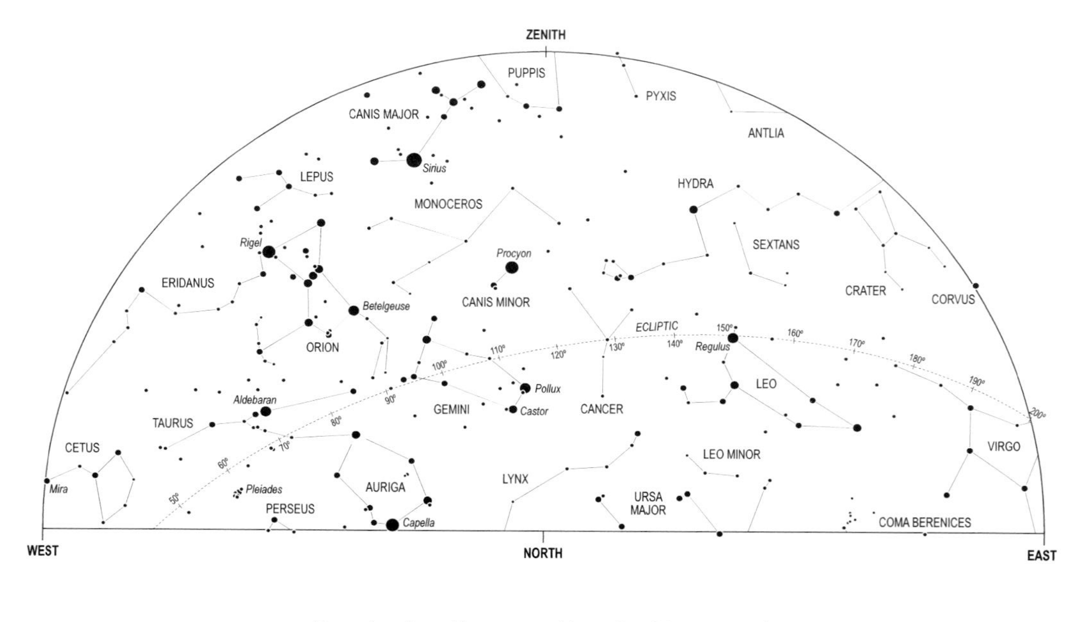

November 6 at 5h	November 21 at 4h
December 6 at 3h	December 21 at 2h
January 6 at 1h	January 21 at midnight
February 6 at 23h	February 21 at 22h
March 6 at 21h	March 21 at 20h

2S

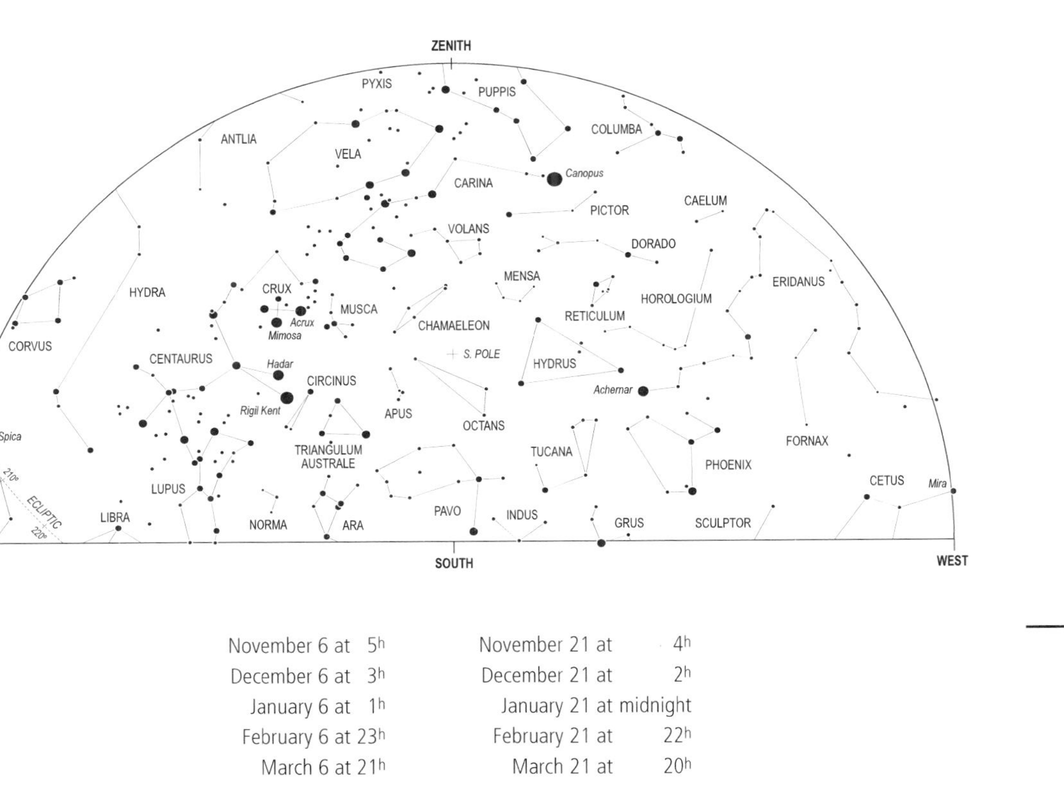

November 6 at 5h	November 21 at 4h
December 6 at 3h	December 21 at 2h
January 6 at 1h	January 21 at midnight
February 6 at 23h	February 21 at 22h
March 6 at 21h	March 21 at 20h

3N

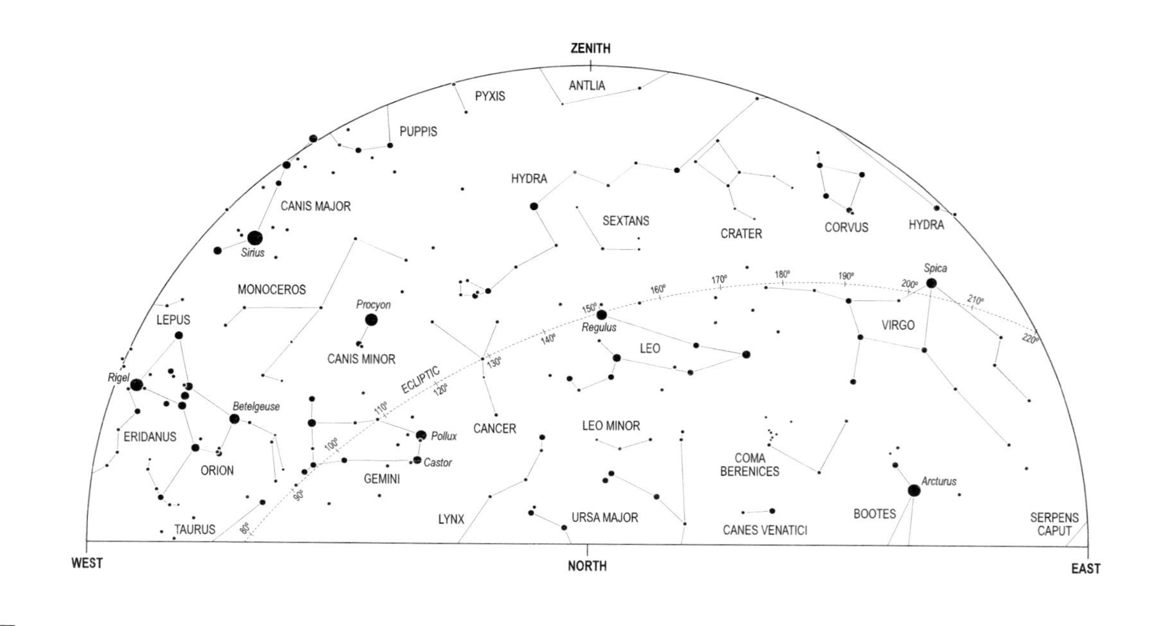

January 6 at 3h	January 21 at 2h
February 6 at 1h	February 21 at midnight
March 6 at 23h	March 21 at 22h
April 6 at 21h	April 21 at 20h
May 6 at 19h	May 21 at 18h

3S

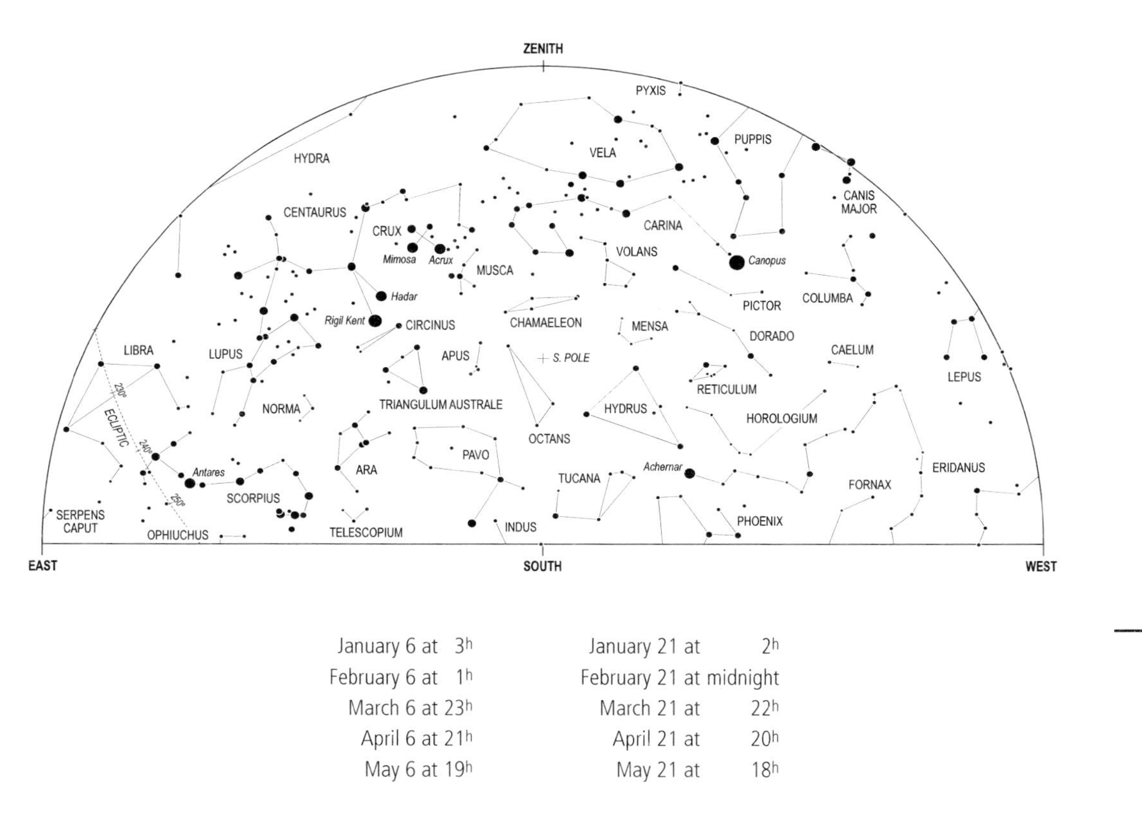

January 6 at 3h	January 21 at 2h
February 6 at 1h	February 21 at midnight
March 6 at 23h	March 21 at 22h
April 6 at 21h	April 21 at 20h
May 6 at 19h	May 21 at 18h

4N

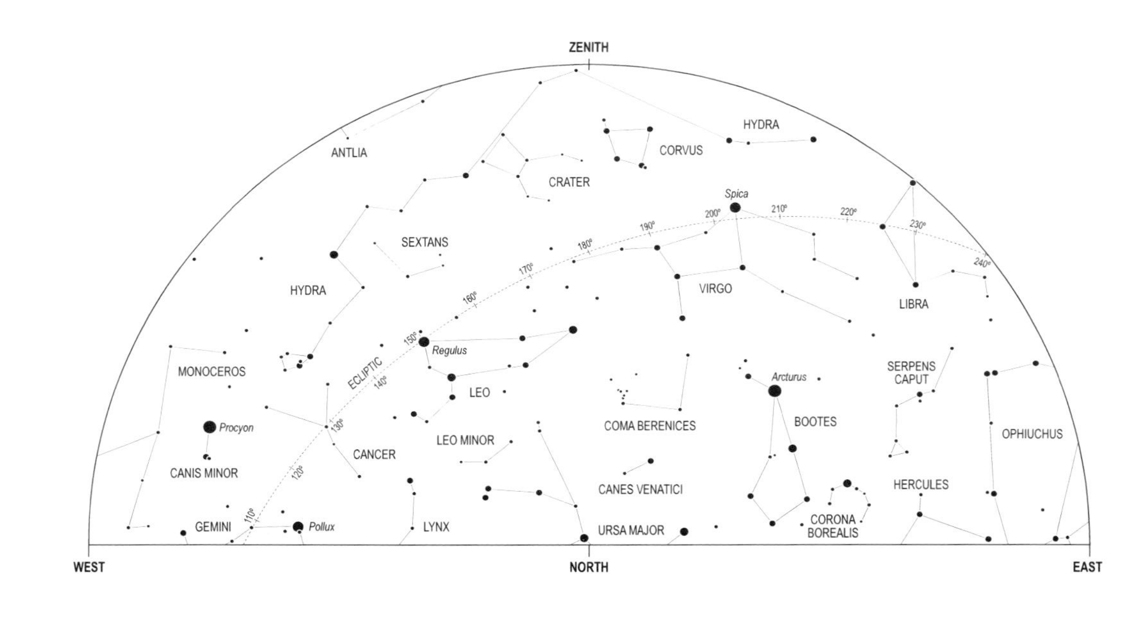

February 6 at 3h
March 6 at 1h
April 6 at 23h
May 6 at 21h
June 6 at 19h

February 21 at 2h
March 21 at midnight
April 21 at 22h
May 21 at 20h
June 21 at 18h

4S

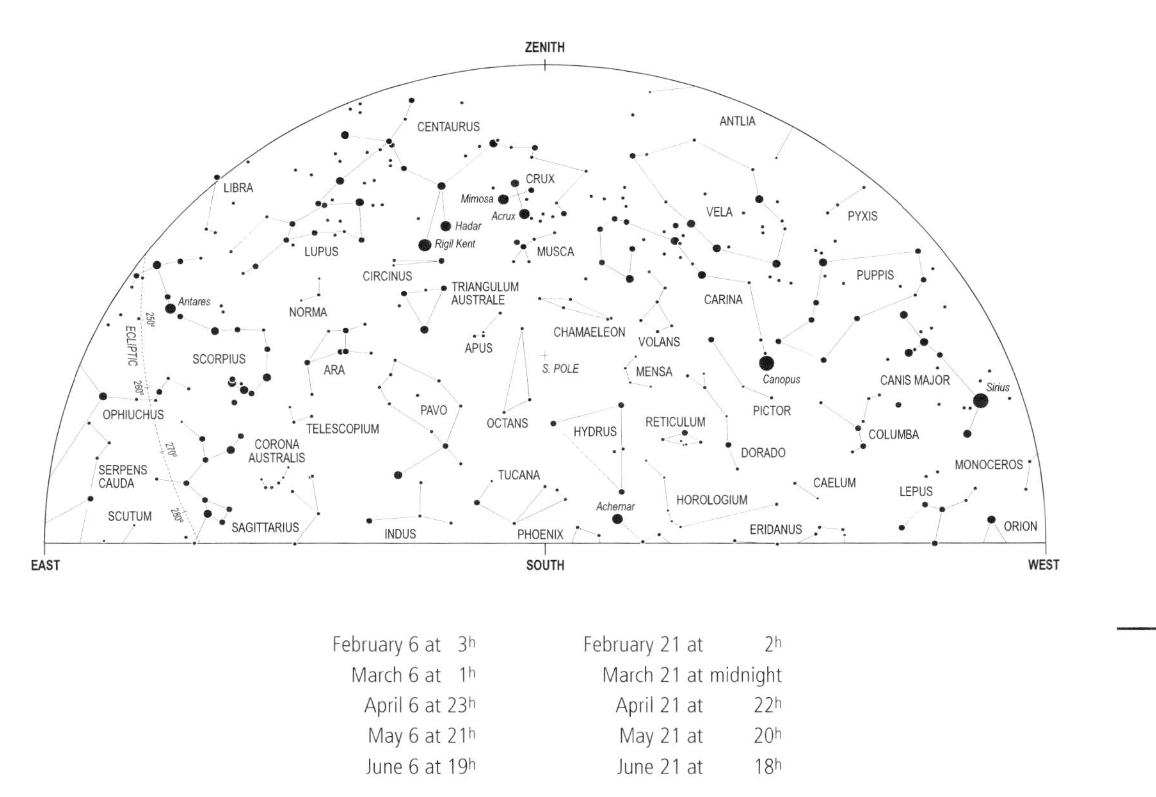

February 6 at 3h	February 21 at 2h
March 6 at 1h	March 21 at midnight
April 6 at 23h	April 21 at 22h
May 6 at 21h	May 21 at 20h
June 6 at 19h	June 21 at 18h

5N

March 6 at 3h	March 21 at 2h
April 6 at 1h	April 21 at midnight
May 6 at 23h	May 21 at 22h
June 6 at 21h	June 21 at 20h
July 6 at 19h	July 21 at 18h

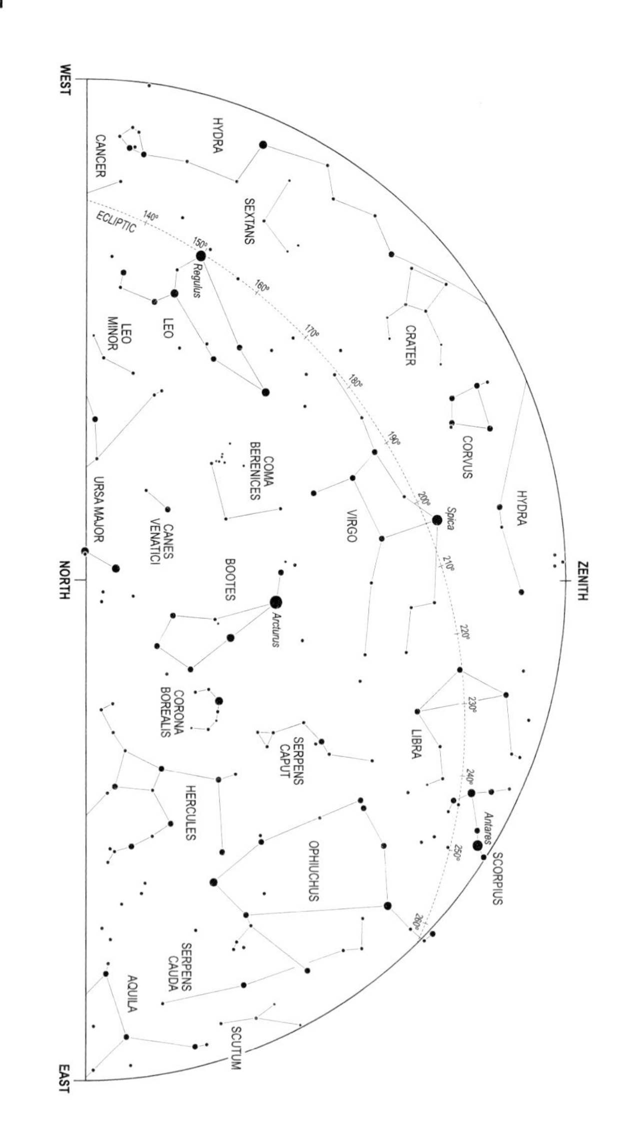

5S

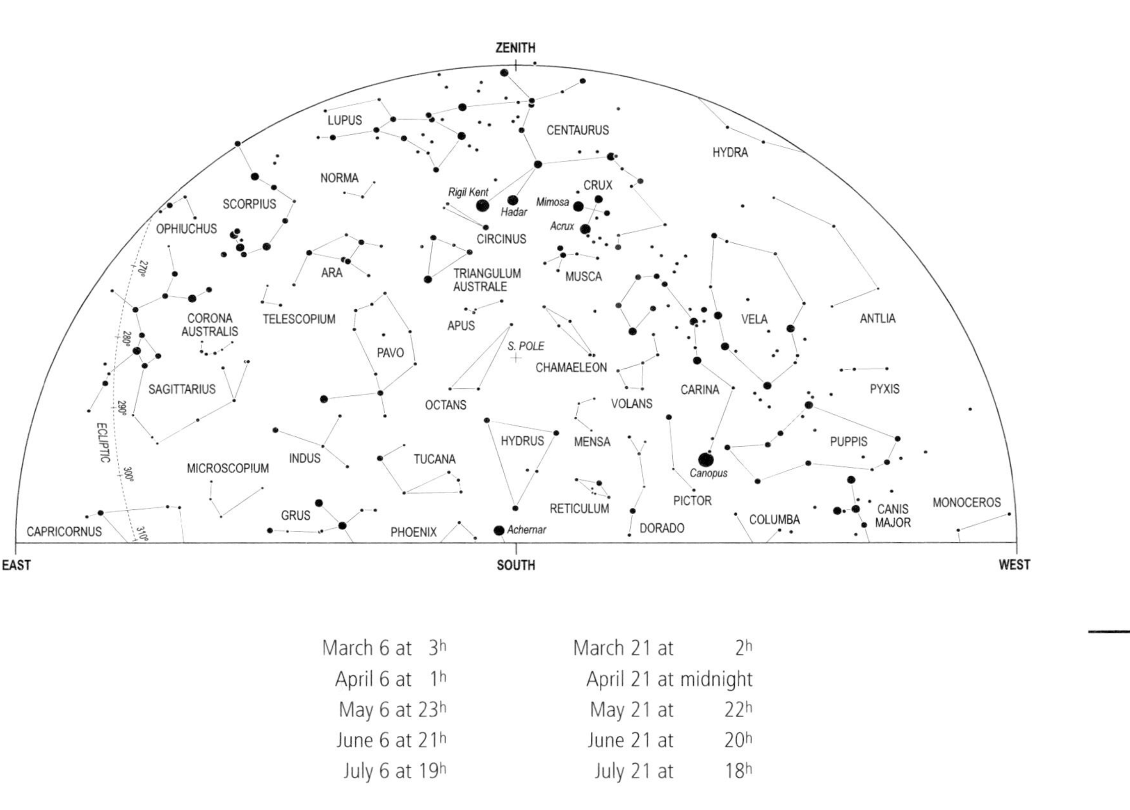

March 6 at 3h
April 6 at 1h
May 6 at 23h
June 6 at 21h
July 6 at 19h

March 21 at 2h
April 21 at midnight
May 21 at 22h
June 21 at 20h
July 21 at 18h

6N

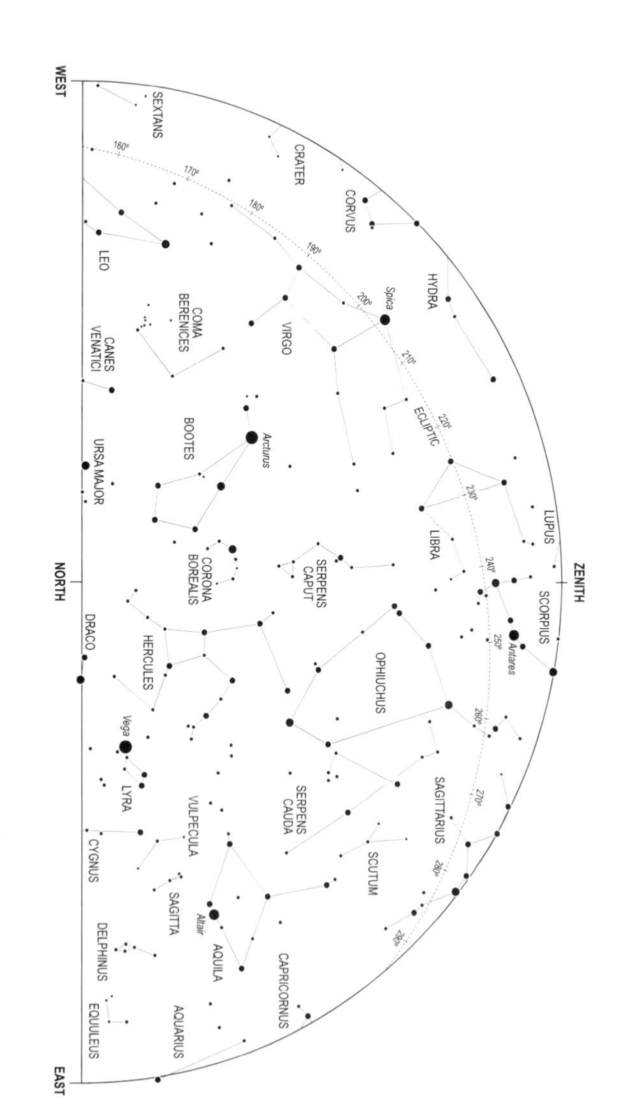

March 6 at 5h	March 21 at 4h
April 6 at 3h	April 21 at 2h
May 6 at 1h	May 21 at midnight
June 6 at 23h	June 21 at 22h
July 6 at 21h	July 21 at 20h

6S

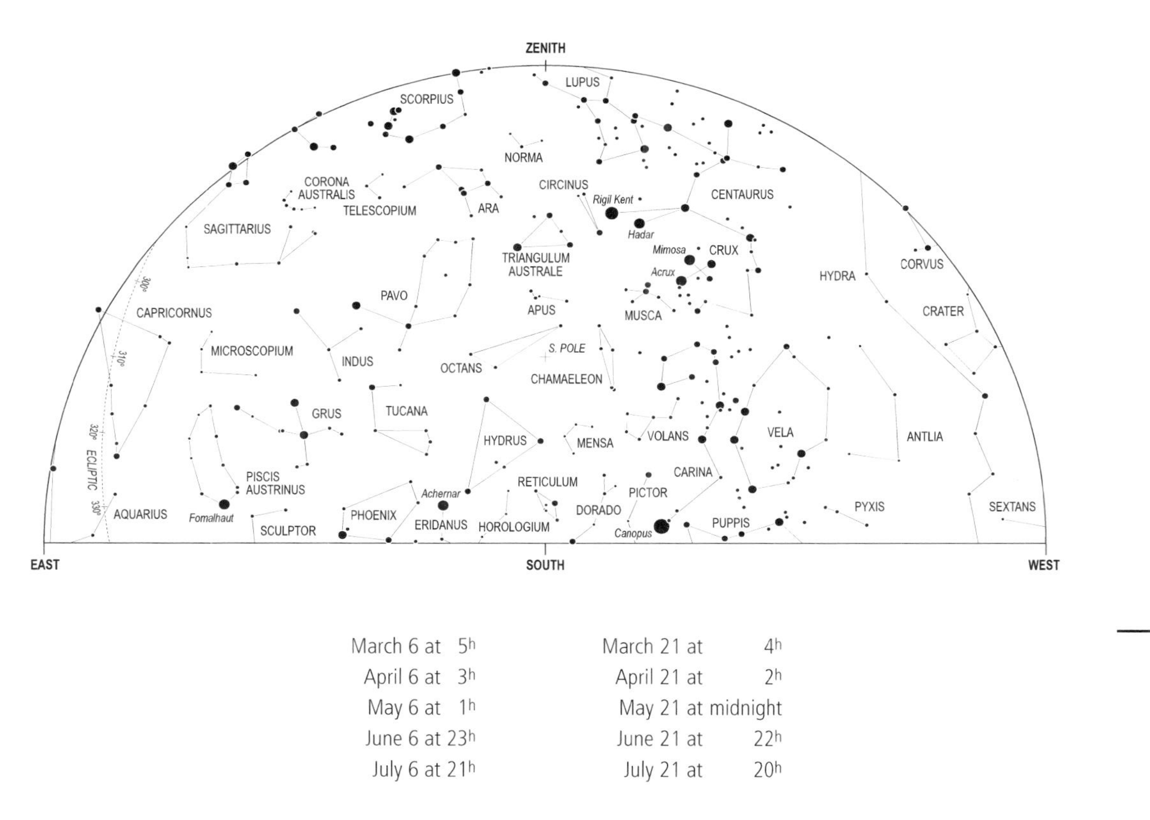

March 6 at 5h	March 21 at 4h
April 6 at 3h	April 21 at 2h
May 6 at 1h	May 21 at midnight
June 6 at 23h	June 21 at 22h
July 6 at 21h	July 21 at 20h

7N

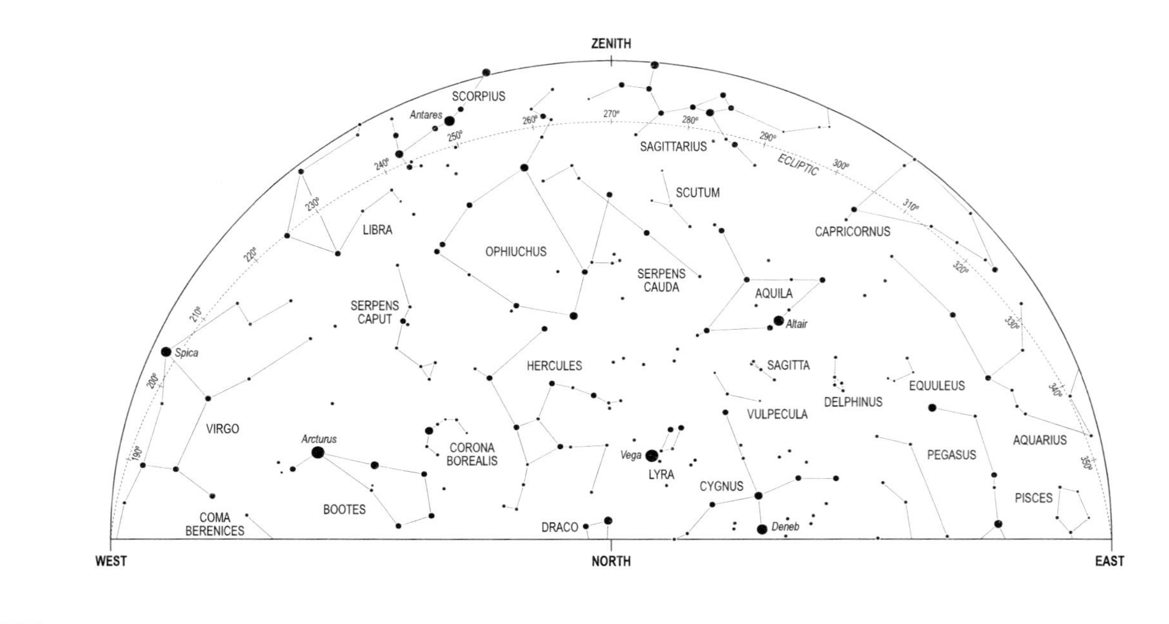

April 6 at 5h	April 21 at 4h
May 6 at 3h	May 21 at 2h
June 6 at 1h	June 21 at midnight
July 6 at 23h	July 21 at 22h
August 6 at 21h	August 21 at 20h

7S

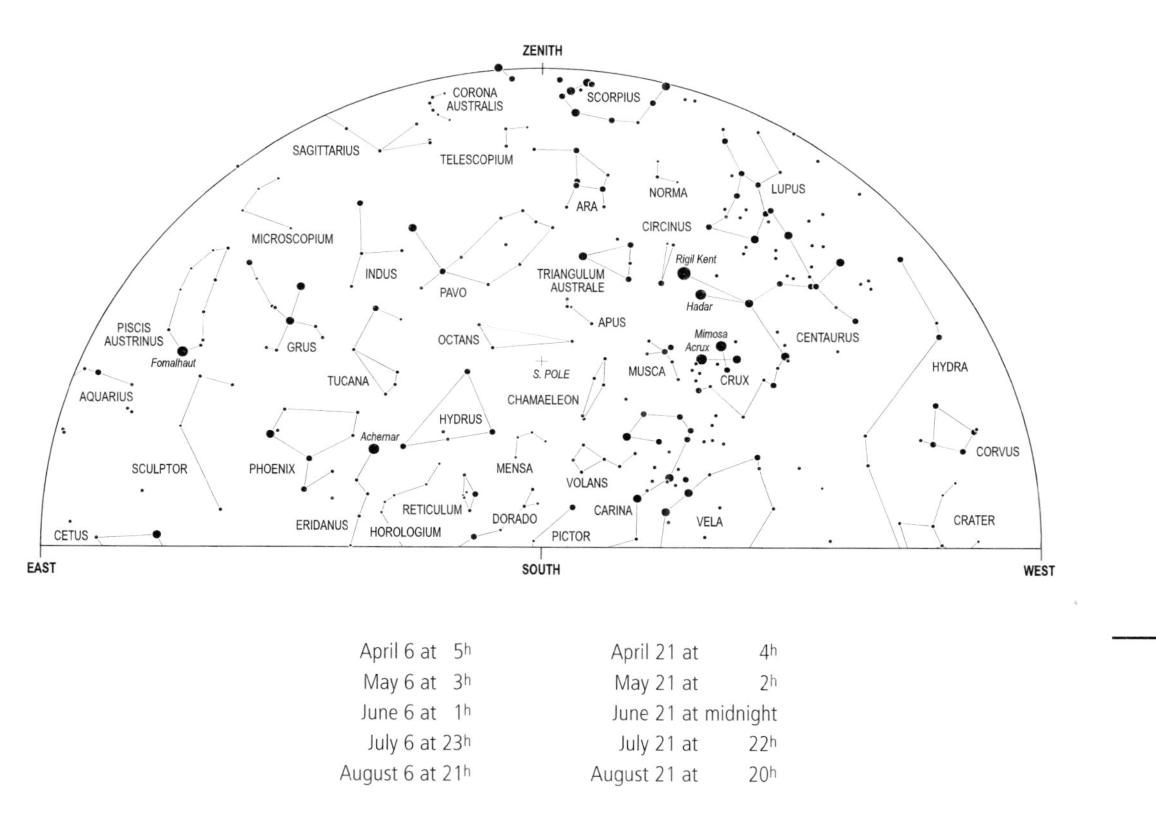

April 6 at 5h	April 21 at 4h
May 6 at 3h	May 21 at 2h
June 6 at 1h	June 21 at midnight
July 6 at 23h	July 21 at 22h
August 6 at 21h	August 21 at 20h

8N

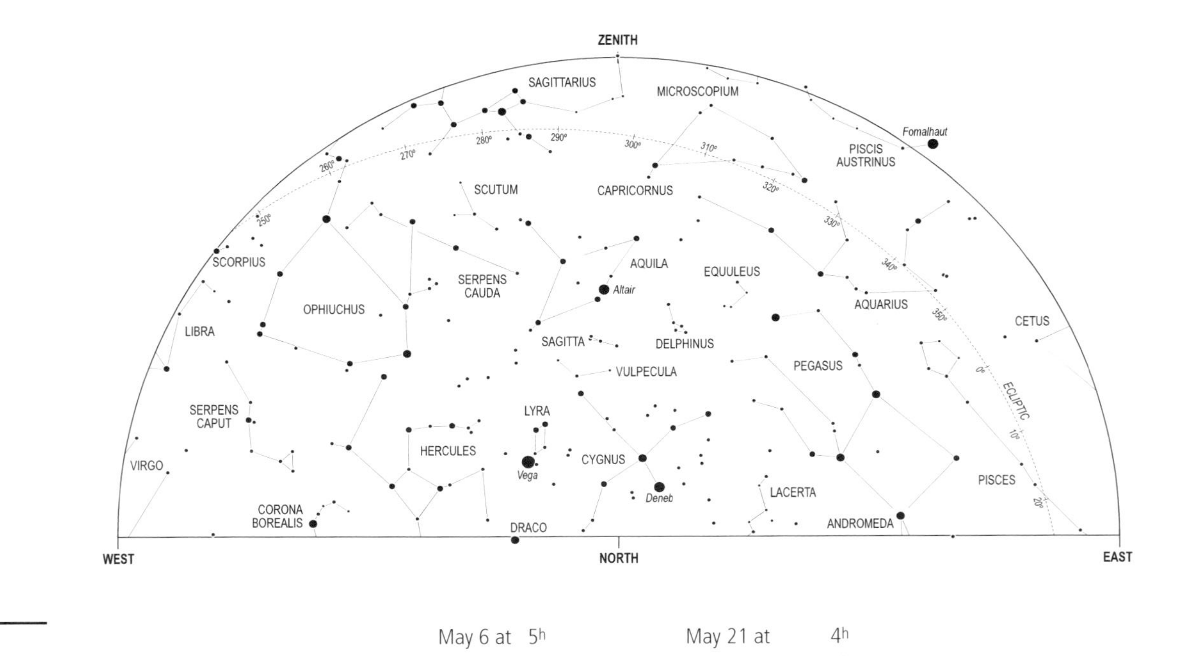

May 6 at 5h	May 21 at 4h
June 6 at 3h	June 21 at 2h
July 6 at 1h	July 21 at midnight
August 6 at 23h	August 21 at 22h
September 6 at 21h	September 21 at 20h

8S

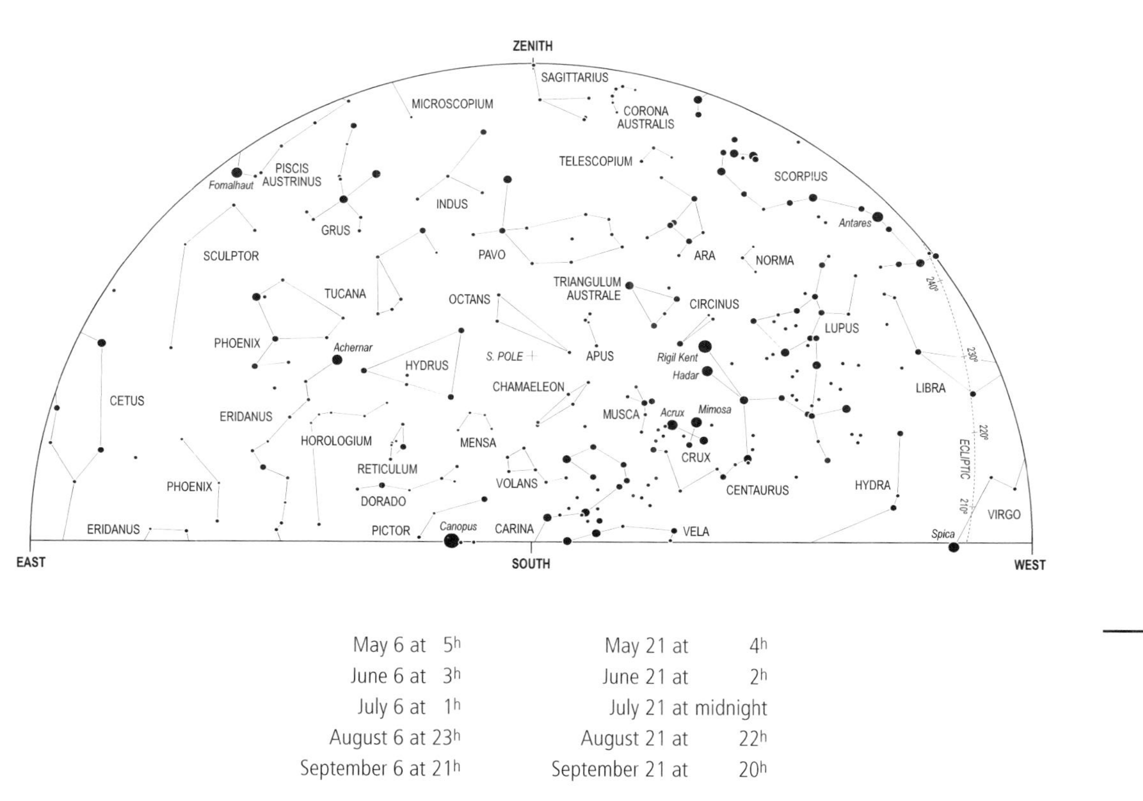

May 6 at 5h
June 6 at 3h
July 6 at 1h
August 6 at 23h
September 6 at 21h

May 21 at 4h
June 21 at 2h
July 21 at midnight
August 21 at 22h
September 21 at 20h

9N

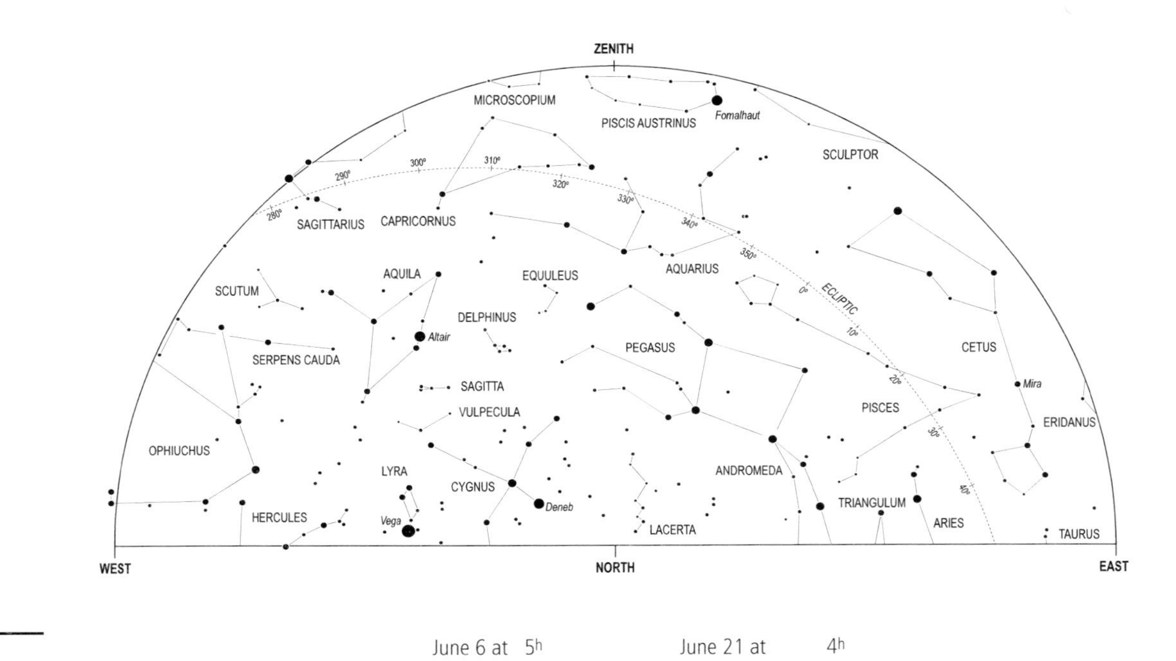

June 6 at 5h
July 6 at 3h
August 6 at 1h
September 6 at 23h
October 6 at 21h

June 21 at 4h
July 21 at 2h
August 21 at midnight
September 21 at 22h
October 21 at 20h

9S

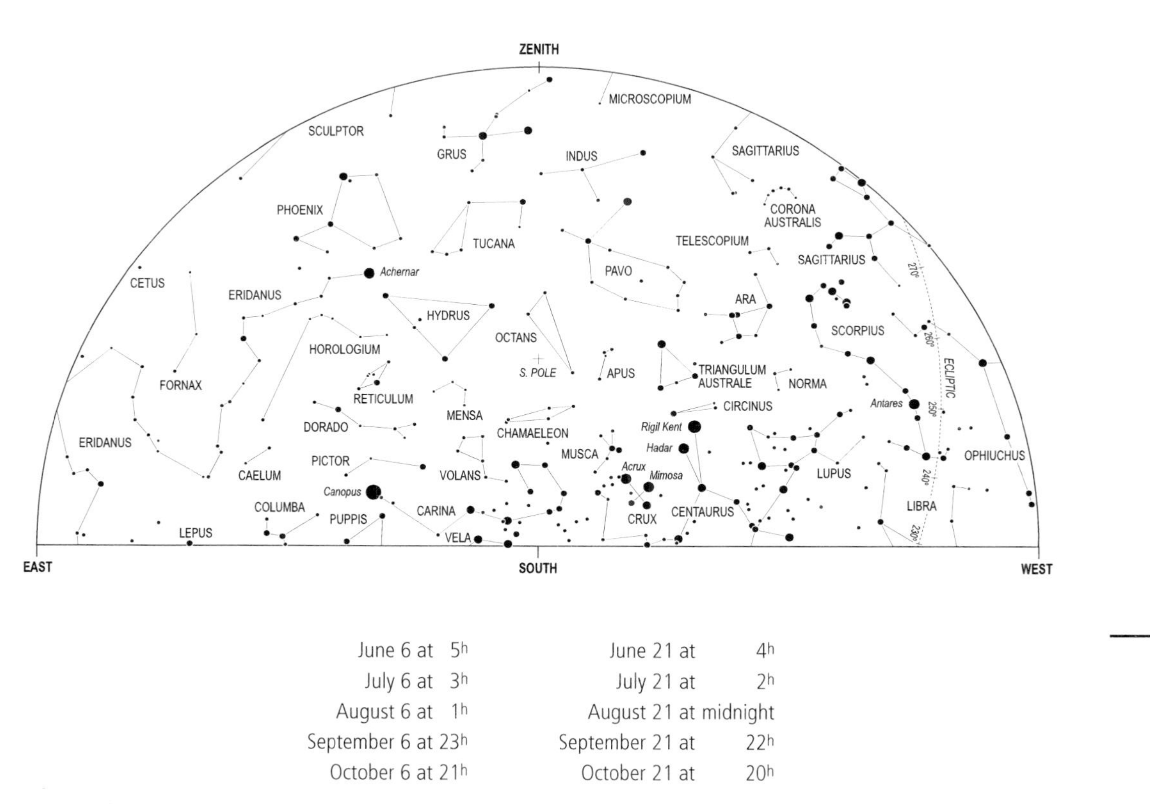

June 6 at 5h	June 21 at 4h
July 6 at 3h	July 21 at 2h
August 6 at 1h	August 21 at midnight
September 6 at 23h	September 21 at 22h
October 6 at 21h	October 21 at 20h

10N

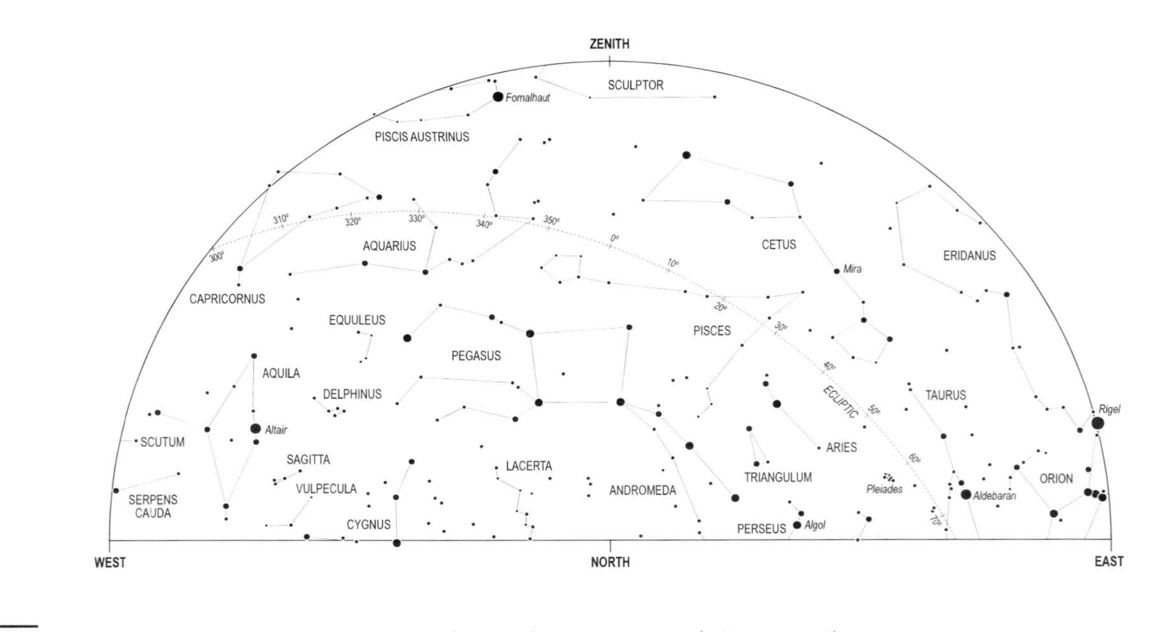

July 6 at 5^h	July 21 at 4^h
August 6 at 3^h	August 21 at 2^h
September 6 at 1^h	September 21 at midnight
October 6 at 23^h	October 21 at 22^h
November 6 at 21^h	November 21 at 20^h

10S

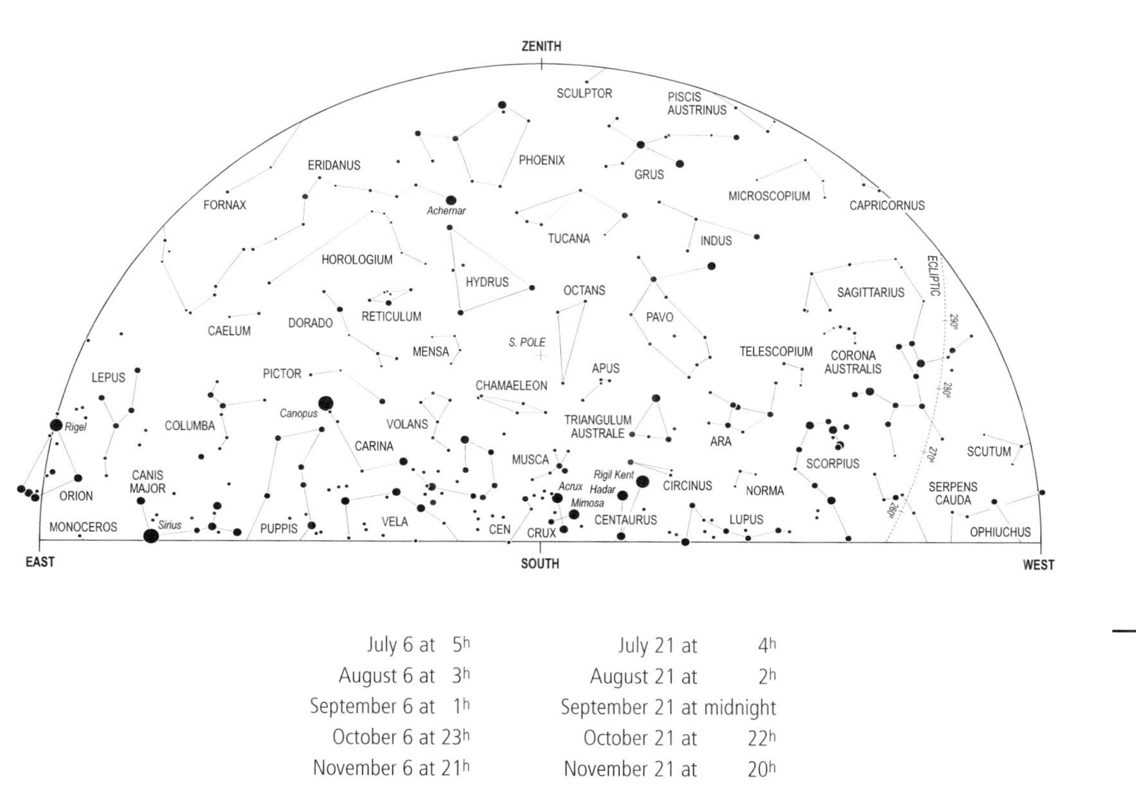

July 6 at 5^h
August 6 at 3^h
September 6 at 1^h
October 6 at 23^h
November 6 at 21^h

July 21 at 4^h
August 21 at 2^h
September 21 at midnight
October 21 at 22^h
November 21 at 20^h

11N

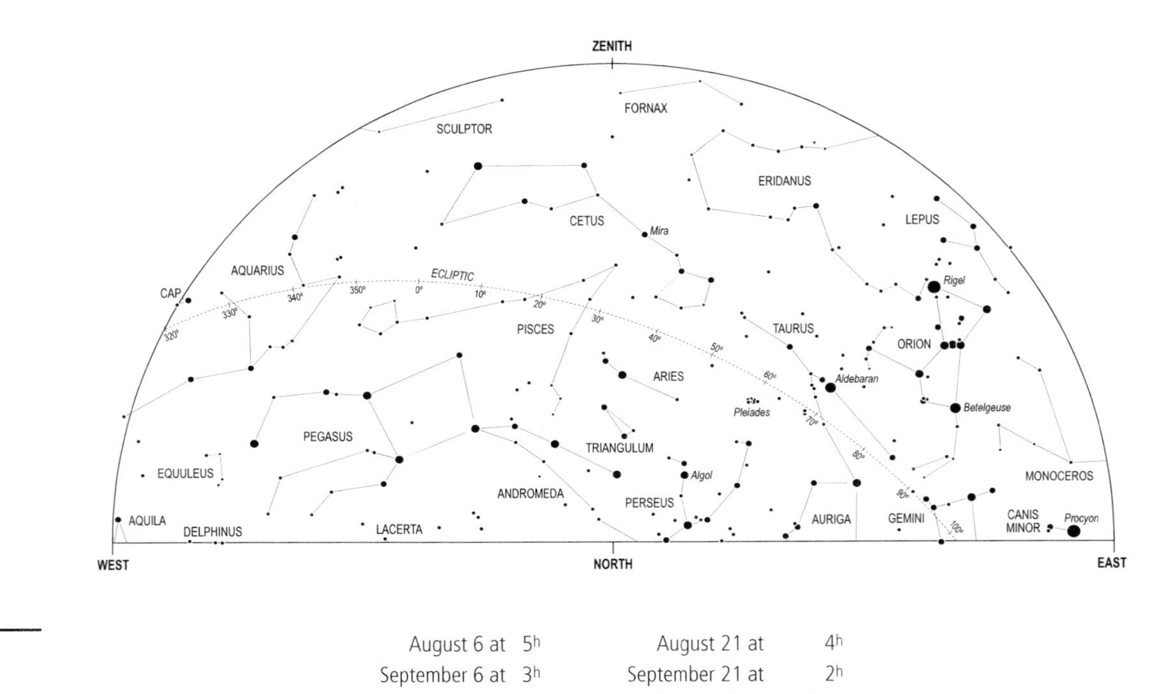

August 6 at 5h	August 21 at 4h
September 6 at 3h	September 21 at 2h
October 6 at 1h	October 21 at midnight
November 6 at 23h	November 21 at 22h
December 6 at 21h	December 21 at 20h

11S

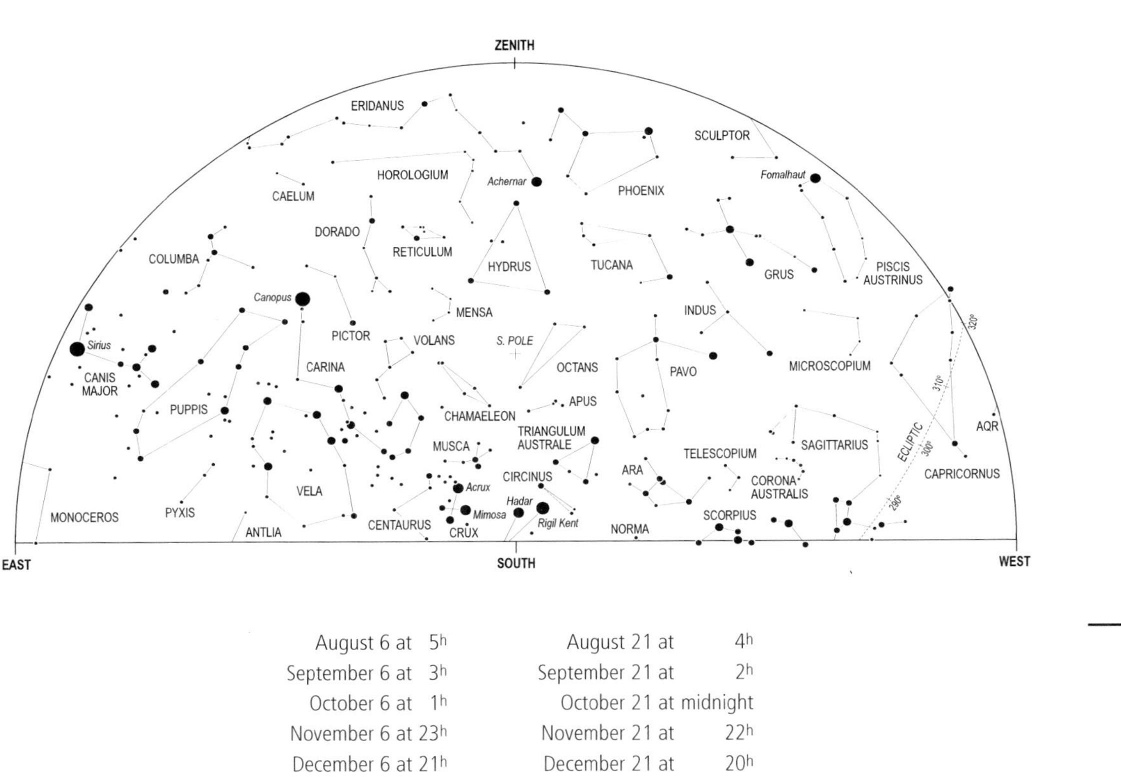

August 6 at 5h	August 21 at 4h
September 6 at 3h	September 21 at 2h
October 6 at 1h	October 21 at midnight
November 6 at 23h	November 21 at 22h
December 6 at 21h	December 21 at 20h

12N

September 6 at 5h
October 6 at 3h
November 6 at 1h
December 6 at 23h
January 6 at 21h

September 21 at 4h
October 21 at 2h
November 21 at midnight
December 21 at 22h
January 21 at 20h

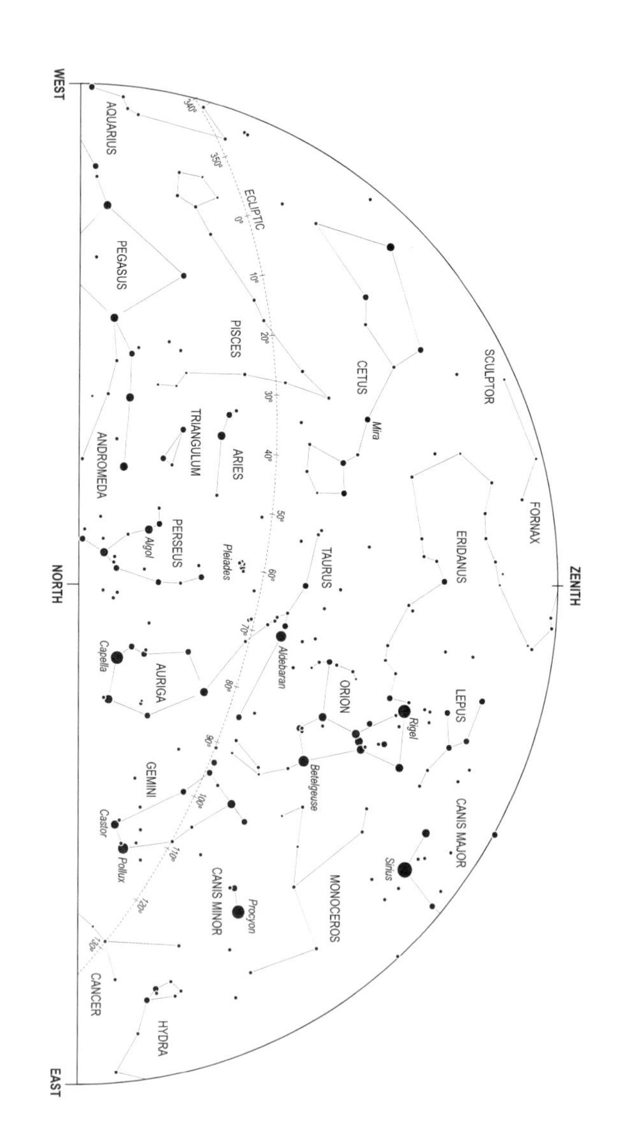

12S

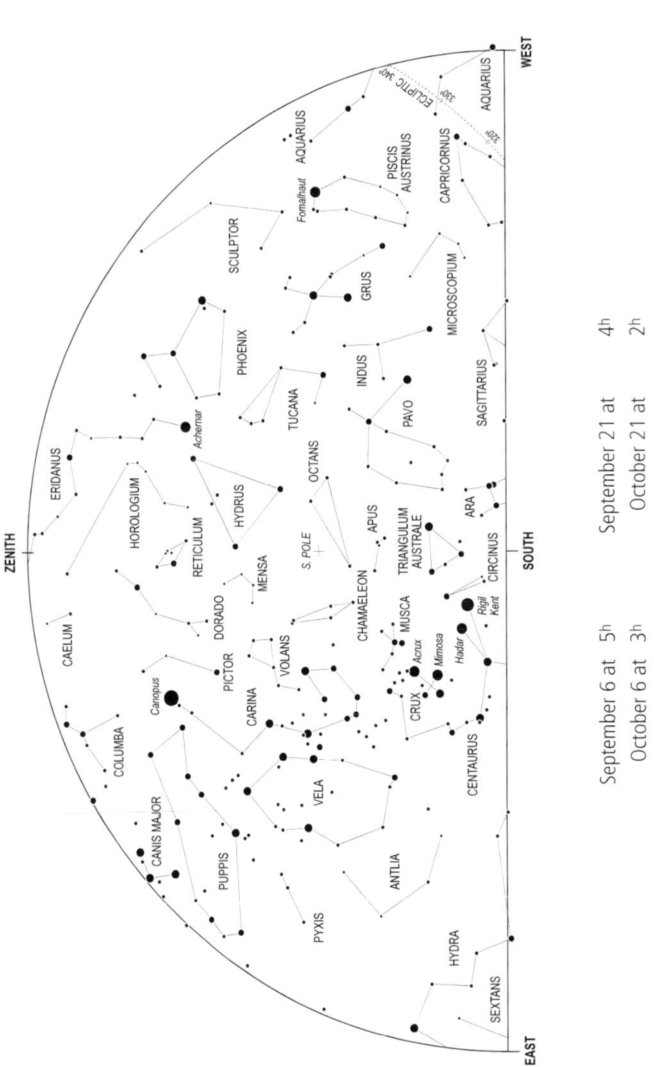

September 6 at 5^h	September 21 at 4^h
October 6 at 3^h	October 21 at 2^h
November 6 at 1^h	November 21 at midnight
December 6 at 23^h	December 21 at 22^h
January 6 at 21^h	January 21 at 20^h

The Planets and the Ecliptic

The paths of the planets about the Sun all lie close to the plane of the ecliptic, which is marked for us in the sky by the apparent path of the Sun among the stars, and is shown on the star charts by a broken line. The Moon and naked-eye planets will always be found close to this line, never departing from it by more than about 7°. Thus the planets are most favourably placed for observation when the ecliptic is well displayed, and this means that it should be as high in the sky as possible. This avoids the difficulty of finding a clear horizon, and also overcomes the problem of atmospheric absorption, which greatly reduces the light of the stars. Thus a star at an altitude of 10° suffers a loss of 60 per cent of its light, which corresponds to a whole magnitude; at an altitude of only 4°, the loss may amount to two magnitudes.

The position of the ecliptic in the sky is therefore of great importance, and since it is tilted at about 23½° to the equator, it is only at certain times of the day or year that it is displayed to the best advantage. It will be realized that the Sun (and therefore the ecliptic) is at its highest in the sky at noon in midsummer, and at its lowest at noon in midwinter. Allowing for the daily motion of the sky, it follows that the ecliptic is highest at midnight in winter, at sunset in the spring, at noon in summer and at sunrise in the autumn. Hence these are the best times to see the planets. Thus, if Venus is an evening object in the western sky after sunset, it will be seen to best advantage if this occurs in the spring, when the ecliptic is high in the sky and slopes down steeply to the horizon. This means that the planet is not only higher in the sky, but will remain for a much longer period above the horizon. For similar reasons, a morning object will be seen at its best on autumn mornings before sunrise, when the ecliptic is high in the east. The outer planets, which can come to opposition (i.e. opposite the Sun), are best seen when opposition occurs in the winter months, when the ecliptic is high in the sky at midnight.

The seasons are reversed in the Southern hemisphere, spring beginning at the September Equinox, when the Sun crosses the Equator on its way south, summer beginning at the December Solstice, when the

Sun is highest in the southern sky, and so on. Thus, the times when the ecliptic is highest in the sky, and therefore best placed for observing the planets, may be summarized as follows:

	Midnight	**Sunrise**	**Noon**	**Sunset**
Northern latitudes	December	September	June	March
Southern latitudes	June	March	December	September

In addition to the daily rotation of the celestial sphere from east to west, the planets have a motion of their own among the stars. The apparent movement is generally *direct*, i.e. to the east, in the direction of increasing longitude, but for a certain period (which depends on the distance of the planet) this apparent motion is reversed. With the outer planets this *retrograde* motion occurs about the time of opposition. Owing to the different inclination of the orbits of these planets, the actual effect is to cause the apparent path to form a loop, or sometimes an S-shaped curve. The same effect is present in the motion of the inferior planets, Mercury and Venus, but it is not so obvious, since it always occurs at the time of inferior conjunction.

The *inferior planets*, Mercury and Venus, move in smaller orbits than that of the Earth, and so are always seen near the Sun. They are most obvious at the times of greatest angular distance from the Sun (greatest elongation), which may reach 28° for Mercury, and 47° for Venus. They are seen as evening objects in the western sky after sunset (at eastern elongations) or as morning objects in the eastern sky before sunrise (at western elongations). The succession of phenomena, conjunctions and elongations, always follows the same order, but the intervals between them are not equal. Thus, if either planet is moving round the far side of its orbit its motion will be to the east, in the same direction in which the Sun appears to be moving. It therefore takes much longer for the planet to overtake the Sun – that is, to come to superior conjunction – than it does when moving round to inferior conjunction, between Sun and Earth. The intervals given in the table at the top of p. 70 are average values; they remain fairly constant in the case of Venus, which travels in an almost circular orbit. In the case of Mercury, however, conditions vary widely because of the great eccentricity and inclination of the planet's orbit.

			Mercury	Venus
Inferior Conjunction	to	Elongation West	22 days	72 days
Elongation West	to	Superior Conjunction	36 days	220 days
Superior Conjunction	to	Elongation East	35 days	220 days
Elongation East	to	Inferior Conjunction	22 days	72 days

The greatest brilliancy of Venus always occurs about 36 days before or after inferior conjunction. This will be about a month after greatest eastern elongation (as an evening object), or a month before greatest western elongation (as a morning object). No such rule can be given for Mercury, because its distances from the Earth and the Sun can vary over a wide range.

Mercury is not likely to be seen unless a clear horizon is available. It is seldom as much as 10° above the horizon in the twilight sky in northern temperate latitudes, but this figure is often exceeded in the Southern Hemisphere. This favourable condition arises because the maximum elongation of 28° can occur only when the planet is at aphelion (farthest from the Sun), and it then lies well south of the equator. Northern observers must be content with smaller elongations, which may be as little as 18° at perihelion. In general, it may be said that the most favourable times for seeing Mercury as an evening object will be in spring, some days before greatest eastern elongation; in autumn, it may be seen as a morning object some days after greatest western elongation.

Venus is the brightest of the planets and may be seen on occasions in broad daylight. Like Mercury, it is alternately a morning and an evening object, and it will be highest in the sky when it is a morning object in autumn, or an evening object in spring. Venus is to be seen at its best as an evening object in northern latitudes when eastern elongation occurs in June. The planet is then well north of the Sun in the preceding spring months, and is a brilliant object in the evening sky over a long period. In the Southern Hemisphere a November elongation is best. For similar reasons, Venus gives a prolonged display as a morning object in the months following western elongation in October (in northern latitudes) or in June (in the Southern Hemisphere).

The *superior planets*, which travel in orbits larger than that of the Earth, differ from Mercury and Venus in that they can be seen opposite the Sun in the sky. The superior planets are morning objects after conjunction with the Sun, rising earlier each day until they come to opposition. They will then be nearest to the Earth (and therefore at their brightest), and will be on the meridian at midnight, due south in northern latitudes, but due north in the Southern Hemisphere. After opposition they are evening objects, setting earlier each evening until they set in the west with the Sun at the next conjunction. The difference in brightness from one opposition to another is most noticeable in the case of Mars, whose distance from Earth can vary considerably and rapidly. The other superior planets are at such great distances that there is very little change in brightness from one opposition to the next. The effect of altitude is, however, of some importance, for at a December opposition in northern latitudes the planets will be among the stars of Taurus or Gemini, and can then be at an altitude of more than 60° in southern England. At a summer opposition, when the planet is in Sagittarius, it may only rise to about 15° above the southern horizon, and so makes a less impressive appearance. In the Southern Hemisphere the reverse conditions apply, a June opposition being the best, with the planet in Sagittarius at an altitude which can reach 80° above the northern horizon for observers in South Africa.

Mars, whose orbit is appreciably eccentric, comes nearest to the Earth at oppositions at the end of August. It may then be brighter even than Jupiter, but rather low in the sky in Aquarius for northern observers, though very well placed for those in southern latitudes. These favourable oppositions occur every fifteen or seventeen years (e.g. in 1988, 2003, 2018), but in the Northern Hemisphere the planet is probably better seen at oppositions in the autumn or winter months, when it is higher in the sky. Oppositions of Mars occur at an average interval of 780 days, and during this time the planet makes a complete circuit of the sky.

Jupiter is always a bright planet, and comes to opposition a month later each year, having moved, roughly speaking, from one Zodiacal constellation to the next.

Saturn moves much more slowly than Jupiter, and may remain in the same constellation for several years. The brightness of Saturn depends on the aspects of its rings, as well as on the distance from Earth and Sun. The Earth passed through the plane of Saturn's rings in 1995 and 1996, when they appeared edge-on; we shall see them at maximum opening, and Saturn at its brightest, this year, in 2002. The rings will next appear edge-on in 2009.

Uranus, *Neptune* and *Pluto* are hardly likely to attract the attention of observers without adequate instruments.

Phases of the Moon 2002

New Moon	d	h	m	First Quarter	d	h	m	Full Moon	d	h	m	Last Quarter	d	h	m
												Jan.	6	03	55
Jan.	13	13	29	Jan.	21	17	46	Jan.	28	22	50	Feb.	4	13	33
Feb.	12	07	41	Feb.	20	12	02	Feb.	27	09	17	Mar.	6	01	24
Mar.	14	02	02	Mar.	22	02	28	Mar.	28	18	25	Apr.	4	15	29
Apr.	12	19	21	Apr.	20	12	48	Apr.	27	03	00	May	4	07	16
May	12	10	45	May	19	19	42	May	26	11	51	June	3	00	05
June	10	23	46	June	18	00	29	June	24	21	42	July	2	17	19
July	10	10	26	July	17	04	47	July	24	09	07	Aug.	1	10	22
Aug.	8	19	15	Aug.	15	10	12	Aug.	22	22	29	Aug.	31	02	31
Sept.	7	03	10	Sept.	13	18	08	Sept.	21	13	59	Sept.	29	17	03
Oct.	6	11	18	Oct.	13	05	33	Oct.	21	07	20	Oct.	29	05	28
Nov.	4	20	34	Nov.	11	20	52	Nov.	20	01	34	Nov.	27	15	46
Dec.	4	07	34	Dec.	11	15	49	Dec.	19	19	10	Dec.	27	00	31

All times are G.M.T.

Longitudes of the Sun, Moon and Planets in 2002

Date		Sun	Moon	Venus	Mars	Jupiter	Saturn
		°	°	°	°	°	°
January	6	285	193	283	351	100	69
	21	301	23	302	1	98	68
February	6	317	245	322	13	97	68
	21	332	68	341	24	96	68
March	6	345	255	357	33	96	69
	21	0	77	16	44	96	70
April	6	16	301	36	55	98	71
	21	31	127	54	65	99	72
May	6	45	334	72	75	102	74
	21	60	165	91	85	104	76
June	6	75	18	110	96	108	78
	21	89	219	127	105	111	80
July	6	104	50	144	115	114	82
	21	118	257	161	125	117	84
August	6	133	97	179	135	121	85
	21	148	306	194	145	124	87
September	6	163	147	208	155	128	88
	21	178	351	219	164	130	89
October	6	193	186	225	174	133	89
	21	207	24	224	183	135	89
November	6	223	239	215	194	137	88
	21	238	69	210	203	138	88
December	6	254	276	214	213	138	87
	21	269	103	224	222	138	85

Longitude of *Uranus* 326°
Neptune 309°

Moon: Longitude of ascending node
Jan. 1: 86° Dec. 31: 67°

Mercury moves so quickly among the stars that it is not possible to indicate its position on the star charts at convenient intervals. The monthly notes must be consulted for the best times at which the planet may be seen.

The positions of the other planets are given in the table on p. 73. This gives the apparent longitudes on dates which correspond to those of the star charts, and the position of the planet may at once be found near the ecliptic at the given longitude.

EXAMPLE

In the Northern Hemisphere two planets are seen in the south-eastern sky in the evenings, in early February. Identify them.

> The Northern Star Chart 12S shows the southern sky at February 6 at 19h and shows longitudes 340° to 130° along the ecliptic. Reference to the table on page 73 gives the longitude of Jupiter as 97° and that of Saturn as 68°. Thus these planets are to be found in the south-eastern sky and the brighter one, which is the farthest east, is Jupiter.

The positions of the Sun and Moon can be plotted on the star maps in the same manner as for the planets. The average daily motion of the Sun is 1°, and of the Moon 13°. For the Moon an indication of its position relative to the ecliptic may be obtained from a consideration of its longitude relative to that of the ascending node. The latter changes only slowly during the year, as will be seen from the values given on p. 73. Let us denote by d the difference in longitude between the Moon and its ascending node. Then if $d = 0°$, 180° or 360°, the Moon is on the ecliptic. If $d = 90°$ the Moon is 5° north of the ecliptic, and if $d = 270°$ the Moon is 5° south of the ecliptic.

On September 21 the Moon's longitude is given in the table on p. 73 as 351° and the longitude of the node is found by interpolation to be about 72°. Thus $d = 279°$ and the Moon is about 5° south of the ecliptic. Its position may be plotted on northern star charts 8S, 9S, 10S, 11S, and 12S, and on southern star charts 8N, 9N, 10N, 11N and 12N.

Some Events in 2002

ECLIPSES

There will be two eclipses, both of the Sun.

June 10–11: annular eclipse of the Sun – Asia, northern Australia, North America.
December 4: total eclipse of the Sun – Africa, Australia.

THE PLANETS

Mercury may be seen more easily from northern latitudes in the evenings about the time of greatest eastern elongation (May 4) and in the mornings around greatest western elongation (October 13). In the Southern Hemisphere the corresponding most favourable dates are around February 21 (mornings) and September 1 (evenings).

Venus is visible in the evenings from late February to October, and in the mornings from November onwards.

Mars does not come to opposition in 2002.

Jupiter is at opposition on January 1 in Gemini.

Saturn is at opposition on December 17 in Taurus.

Uranus is at opposition on August 20 on the Capricornus–Aquarius border.

Neptune is at opposition on August 2 in Capricornus.

Pluto is at opposition on June 7 in Ophiuchus.

Monthly Notes 2002

January

New Moon: January 13 *Full Moon*: January 28

EARTH is at perihelion (nearest to the Sun) on January 2, at a distance of 147 million kilometres (91.4 million miles).

MERCURY attains its greatest eastern elongation (19°) from the Sun on January 11. For observers in northern and equatorial latitudes it is visible as an evening object for the first three weeks of the month, though for observers in southern latitudes the period of visibility is reduced to the first half of the month. This elusive planet may be glimpsed low above the south-western horizon at the end of evening civil twilight. At the beginning of January Mercury has a magnitude of −0.7 but this has faded to +0.8 three weeks later. Mercury is unobservable later in the month as it passes through inferior conjunction on January 27.

VENUS passes slowly through superior conjunction on January 14 and therefore remains unsuitably placed for observation throughout the month.

MARS is visible in the south-western sky in the evenings, magnitude +0.9. The planet is moving eastwards, south of the Square of Pegasus, as seen in Figure 6 (given with the notes for May) which shows the path of Mars among the stars.

JUPITER is at opposition on the first day of the year and is therefore visible throughout the hours of darkness. Jupiter now has a magnitude of −2.7 and is retrograding slowly in the western part of the constellation of Gemini, as will be seen by reference to Figure 1 which shows the path of the planet among the stars during the year. At opposition Jupiter is 626 million kilometres (389 million miles) from the Earth.

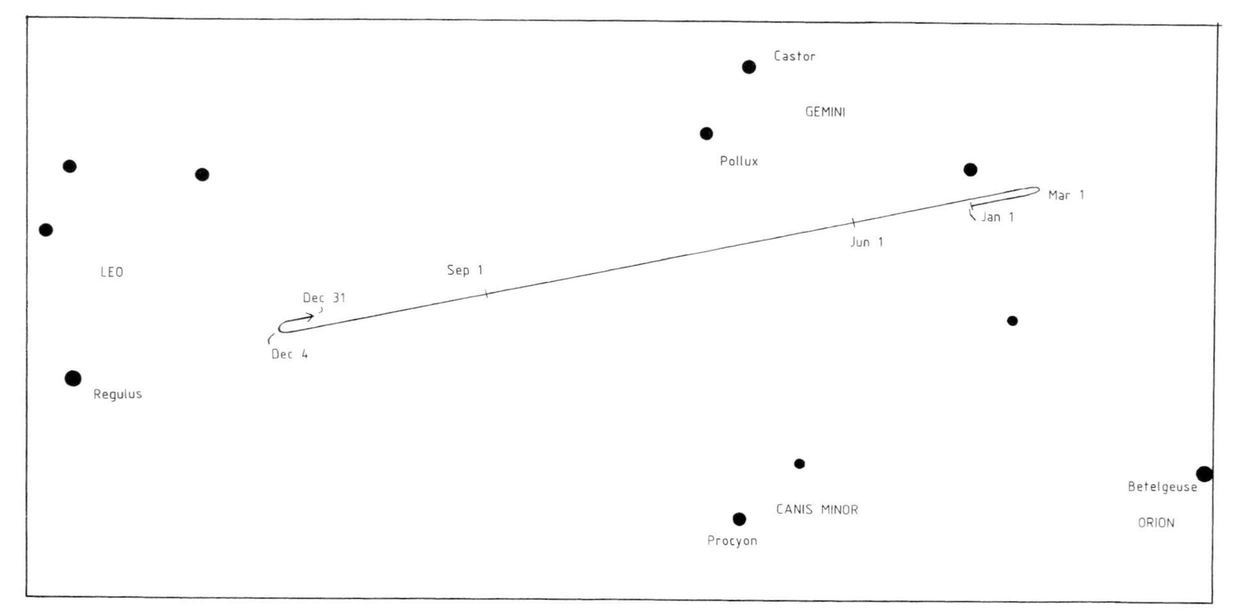

Figure 1. The path of Jupiter during 2002.

SATURN is an evening object, magnitude −0.2. Saturn was at opposition last month and therefore is now visible for the greater part of the hours of darkness. The planet is moving slowly westwards in the constellation of Taurus, north of the Hyades. Its path among the stars is shown in Figure 4, given with the notes for March.

Oppositions of the Planets. During 2002 there are oppositions of Jupiter, Saturn, Uranus, Neptune and Pluto as indeed there are in almost every year. With Mars, the situation is different. There has been no opposition since June 13, 2001, and the next will not occur until August 28, 2003, because the synodic period of Mars – 780 days – is much longer than that of any other planet.

The synodic period is the mean interval between successive oppositions. To explain what is meant, let us first consider Pluto, which moves round the Sun at a mean distance of more than 5,900 million kilometres (though its orbit is unusually eccentric, and at its closest it can come within the orbit of Neptune). As well as having a large orbit, Pluto is slow-moving, taking nearly 248 years to complete one revolution around the Sun. In one year, the time taken for the Earth to go once round the Sun, Pluto covers only a tiny fraction of its orbit – so that the Sun, Earth and Pluto are lined up every 366.7 days. Having been right round the Sun in 365.25 days, the Earth takes only about an extra day and a half to catch up with Pluto.

With the closer planets, more time is required; the synodic periods for Neptune, Uranus, Saturn and Jupiter are respectively 367.5, 369.7, 378.1 and 398.9 days – so that, for instance, oppositions of Jupiter occur on average about 34 days later in each year; the opposition of January 1, 2002 follows that of November 28, 2000, and there will also be oppositions on February 2, 2003, March 4, 2004, and so on.

Mars is a special case. Its mean orbital velocity is comparable with that of the Earth, and the synodic period is therefore a great deal longer. Moreover, the greater eccentricity of Mars' orbit leads to considerable variation in the interval between successive oppositions; although the synodic period is 780 days, the interval may be as short as 764 days or as long as 810 days. Consequently, oppositions of Mars do not occur every year; thus there were oppositions in 1997, 1999 and 2001, but there will be none in 2002 – and following the opposition of 2003, 2004 will be another 'blank year' so far as Mars is concerned.

Of course, Mars will not be invisible during 2002. In January it is visible in the south-western sky in the evenings, and by December it will be visible for several hours in the south-eastern sky before dawn. However, it will be a long way away, and not even large telescopes will show much upon its disk. It must be admitted that observers of Mars must resign themselves to inactivity through most of 2002.

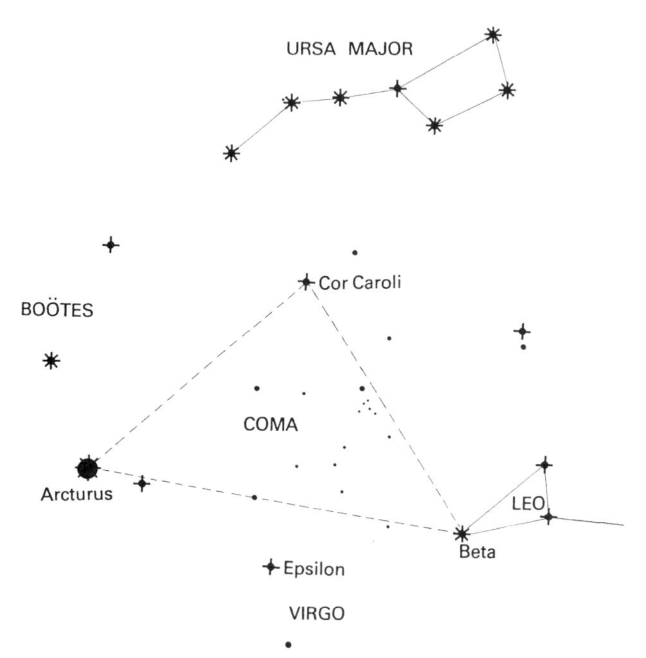

Figure 2. How to find the star Cor Caroli using Beta Leonis (Denebola) in the tail of the Lion, Alkaid, the last star in the handle of the Plough and Arcturus in Boötes.

Cor Caroli. On a January evening at about 2200 hours, Leo the Lion and Ursa Major, the Great Bear, are well displayed in the north-eastern sky. A line from Denebola (the tail of the Lion) to Alkaid (the last star in the tail of the Bear, or the handle of the Plough) may be seen as being divided into three roughly equal parts by the star cluster of Coma Berenices and the star Cor Caroli. This is a fine double star, the brightest in the constellation Canes Venatici, the Hunting Dogs, and its name (Charles' Heart) is in remembrance of King Charles II. The story goes that on the evening before the return of Charles to London, the court physician, Scarborough, remarked that this star shone with a peculiar lustre, and suggested that it should be named in honour of the king. It is certainly true that many old star atlases (even those of a much later date) marked the star with a picture of a heart surmounted with a crown. The star actually had its place on the collar of Chara, one of the hunting dogs, but the incongruity of the crowned heart did not seem to matter! In later atlases and globes, the royal heart disappeared, but it is perhaps fitting that the king who founded the Royal Society and the Royal Observatory should be remembered in this simple way.

February

New Moon: February 12 *Full Moon*: February 27

MERCURY, although it reaches greatest western elongation (27°) on February 21, is not suitably placed for observation from the latitudes of the British Isles. For observers further south this will be the most favourable morning apparition of the year. Figure 3 shows, for observers in latitude 35° S, the changes in azimuth (true bearing from the north through east, south and west) and altitude of Mercury on successive evenings when the Sun is 6° below the horizon. This condition is known as the beginning of morning civil twilight and in this latitude and at this time of year occurs about 30 minutes before sunrise. The changes in the brightness of the planet are indicated by the relative sizes of the circles marking Mercury's position at five-day intervals. During February its magnitude brightens from +0.8 to 0.0. It will be noticed that Mercury is at its brightest after it reaches greatest western elongation.

VENUS is too close to the Sun for observation during the first half of the month, but then becomes visible as an evening object, magnitude −3.9, low above the west-south-western horizon for a short while after sunset.

MARS continues to be visible as an evening object in the south-western sky, magnitude +1.2. At the beginning of February, Mars is in the constellation of Pisces, but enters Aries at the very end of the month.

JUPITER, magnitude −2.5, continues to be visible as a conspicuous evening object in the southern skies. Jupiter is almost stationary in the western part of Gemini. The four Galilean satellites are readily observable with a small telescope or even a good pair of binoculars provided that they are held rigidly.

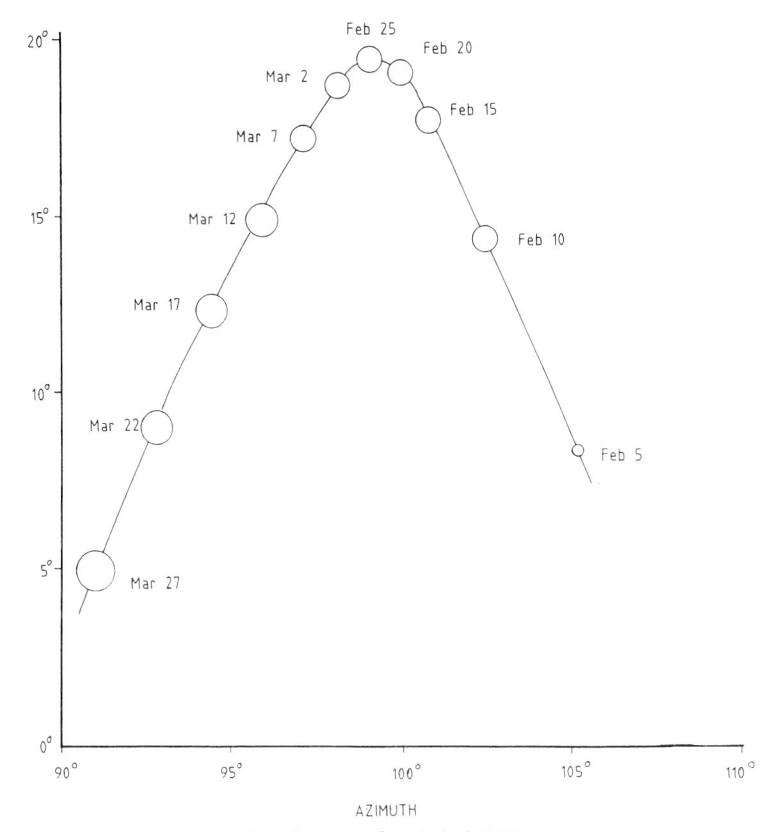

Figure 3. Morning apparition of Mercury, from latitude 35°S.

SATURN continues to be visible in the evening sky, magnitude 0.0. On February 8 it reaches its second stationary point in the constellation of Taurus, thereafter resuming its direct motion.

The Fading of Mars. Having approached the Earth to within 67 million kilometres (42 million miles) in June 2001, when it was a brilliant object of magnitude −2.4 in Ophiuchus, Mars has now drawn away from us, and is receding all the time. By the end of February it will have faded to below first magnitude (slightly inferior to Aldebaran), whereas in June last year it was the brightest object in the entire sky apart from the Sun, Moon and Venus.

Because it is a small body and its distance from us is so variable, Mars undergoes changes in apparent brightness which are more noticeable than with any other planet. Venus and Jupiter are always strikingly brilliant when they are visible at all, and Saturn remains between magnitudes 0 and 1 almost all the time. Mars, however, can be either exceptionally bright or else relatively obscure; at its dimmest it is comparable with Polaris, and unwary observers can all too easily mistake it for an ordinary red star. For most of 2002, Mars will not be particularly prominent, becoming steadily less and less evident until it reaches conjunction in August. It will then reappear as a morning object, but even by the end of the year it will still be well below first magnitude.

Of course, adequate telescopes will show markings on the small Martian disk; but modest instruments will not show very much, and for most amateurs Mars may be to all intents and purposes disregarded throughout much of 2002.

Mars next comes to opposition in August 2003, when it will be only 56 million kilometres (35 million miles) distant – rather closer than it was in June 2001 – and even more brilliant at magnitude −2.7. Unfortunately, for observers in Northern Europe and North America, this will be a rather disappointing opposition since Mars will be low in the southern sky among the stars of Capricornus.

Extinction. Early on February evenings, several brilliant stars are well placed. It is interesting to compare Rigel, in the 'foot' of Orion, with Capella, in Auriga. The two are almost exactly equal: Capella is given as magnitude 0.08, and Rigel as 0.12, so Capella is just four-hundredths of a magnitude the brighter, which is negligible by naked-eye standards. Yet from Britain and North America, Capella will look much more brilliant. The reason for this is that during winter evenings, Capella is almost overhead. Rigel, which lies south of the celestial equator, on the other hand, is much lower down, and some of its glory is lost by 'extinction', due to the fact that its light is coming to us obliquely through a much thicker layer of the Earth's atmosphere. A star which is 20° above the horizon will lose almost half a magnitude, while at 1° altitude the loss is as much as three magnitudes. Only above 45° altitude may extinction be neglected for practical purposes. A comparison between Rigel and Capella will show just how marked the extinction effect is.

March

New Moon: March 14 *Full Moon*: March 28

Equinox: March 20

Summer Time in the United Kingdom commences on March 31.

MERCURY, for observers in the British Isles, remains too close to the Sun for observation throughout the month. For observers further south, Mercury continues to be visible as a morning object for the first three weeks of the month, as shown on Figure 3 (given with the notes for February). During this period its magnitude brightens from 0.0 to −0.9.

VENUS is visible as an evening object, magnitude −3.9, low in the western sky after sunset.

MARS, magnitude +1.4, continues to be visible in the south-western sky in the early evenings. The slightly reddish appearance of the planet is an aid to its identification.

JUPITER continues to be visible as a conspicuous object in the evening skies from shortly after sunset until well after midnight. Its magnitude is −2.3. On the first day of the month Jupiter reaches its second stationary point in the western part of Gemini and afterwards resumes its direct motion.

SATURN, magnitude +0.1, is still an evening object in the western sky. Before the end of the month even the more favourably placed observers (those in northern temperate latitudes), will have noticed that it has sunk below the horizon before midnight. Figure 4 shows the path of Saturn among the stars during 2002. The rings of Saturn present a wonderful spectacle to the observer with a small telescope. In 1995 the Earth passed through Saturn's ring plane on two occasions, and since then the rings have been slowly opening up: in 2002 the diameter of the minor axis has now reached 20 arc seconds, which is just greater than the polar diameter of the planet itself. The rings will not be at their maximum opening until early in 2003, when Saturn will be an evening object.

The First Point of Aries. On March 20, the Sun crosses the celestial equator, moving from south to north, after which date it will stay in the northern part of the sky until September 23. March 20 is therefore the date of the spring or vernal equinox (in the northern hemisphere) this year, and the Sun will lie at what is known as the First Point of Aries.

The positions of celestial bodies are given in terms of declination and right ascension. Declination is simply the angular distance north or south of the celestial equator, so that it corresponds to latitude on the Earth's surface (though, rather confusingly, the term 'celestial latitude' has a rather different meaning). The declination of the north celestial

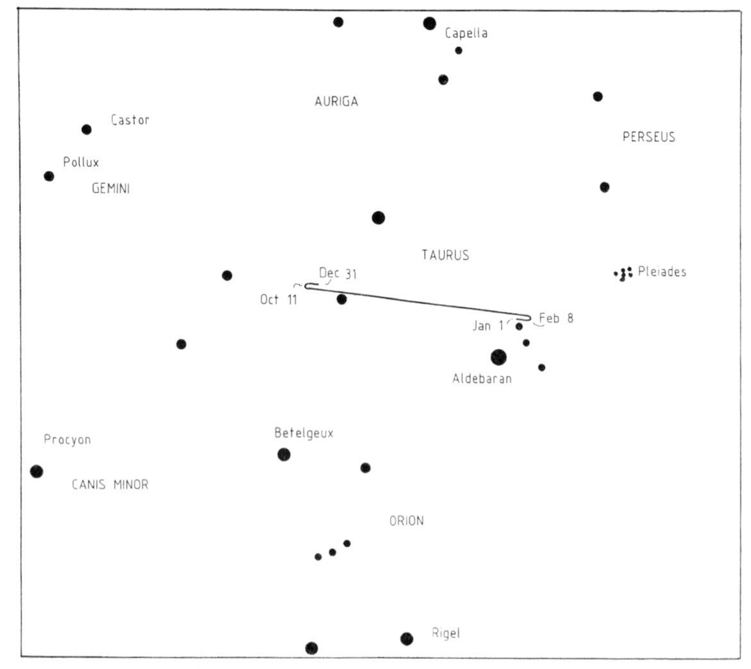

Figure 4. The path of Saturn during 2002.

pole is, of course, 90°N, and the current pole star, Polaris, is within 1° of this. Delta Orionis, the northernmost star in the Belt of Orion, the Hunter, is very close to the celestial equator.

Right ascension is measured eastwards from the First Point of Aries. This is simply the place in the sky where the ecliptic (which may be defined as the Sun's apparent yearly path against the background stars) cuts the celestial equator. The right ascension of a body is measured in hours, minutes and seconds of time, and is the interval that elapses between the meridian passage of the First Point of Aries and the meridian passage of the star. Thus Sirius, in Canis Major, reaches the meridian 6 hours 45 minutes after the First Point of Aries has done so; the right ascension of Sirius is therefore 6h 45m.

The Sun, Moon and planets change their right ascensions and declinations comparatively quickly. The stars, which are to all intents and purposes fixed relative to each other, do not shift so rapidly. The only alterations are due to the effects of precession, or the movement of the direction of the Earth's axis. Because of precession, the polar point – and hence the equator – shifts very gradually, and so the position of the First Point of Aries changes too. When the Egyptian Pyramids were built, the pole star was the comparatively obscure Thuban in the constellation of Draco, and in 12,000 years' time, the brilliant star Vega will be the north polar star.

The first Point of Aries is no longer in Aries. Precession over the centuries has shifted it into the adjacent constellation of Pisces, the Fishes; so strictly speaking, Pisces should now be regarded as the first constellation of the Zodiac instead of the last. There is no bright star close to the First Point; the position lies roughly between Iota Piscium (magnitude 4.1) and Iota Ceti (magnitude 3.6), rather closer to Iota Piscium. In the course of time, the First Point will move out of Pisces and into the next zodiacal constellation, Aquarius.

The discovery of Pallas. This month we mark the bicentenary of the discovery of the second asteroid or minor planet, Pallas, by the German astronomer, Heinrich Wilhelm Olbers. Following the discovery of the first asteroid, Ceres, by Giuseppe Piazzi from Palermo, Sicily, on January 1, 1801, there was a danger that it might be lost, because no reliable orbit could be computed with the methods then available. The great German mathematician and Director of the Observatory at Göttingen, Carl Friedrich Gauss, came to the rescue when he invented a powerful new technique for orbital calculation. Using his new method, Gauss computed the orbit of Ceres and predicted where the new minor planet would be during the winter of 1801–2. These predictions were used by Heinrich Olbers and the Hungarian astronomer Baron Franz von Zach to search for Ceres towards the end of 1801, and von Zach recovered it first on December 7. Olbers, a doctor by profession and a skilled amateur astronomer, recovered Ceres himself on January 1, 1802, from his private observatory in Bremen. He continued to monitor the object for several months, and it was during these follow-up observations of Ceres, on March 28, 1802, that Olbers found a second asteroid, which he named Pallas. Later measurements showed that Pallas is the second largest asteroid after Ceres; Pallas is ellipsoidal in shape and measures 571 x 525 x 482 kilometres (355 x 326 x 300 miles). Unfortunately, Pallas is not favourably placed for observation this year. Incidentally, Olbers also discovered the fourth asteroid – Vesta – in 1807.

April

New Moon: April 12 *Full Moon*: April 27

MERCURY is too close to the Sun for observation at first as it passes through superior conjunction on April 7. However, just after the middle of the month it becomes visible in the evenings, low above the west-north-western horizon, for observers in equatorial and northern latitudes. For observers in northern temperate latitudes this will be the most favourable evening apparition of the year. Figure 5 shows, for observers in latitude 52° N, the changes in azimuth (true bearing from the north through east, south and west) and altitude of Mercury on successive evenings when the Sun is 6° below the horizon. This condition is known as the end of evening civil twilight, and in this latitude and at this time of year occurs about 35 minutes after sunset. The changes in the brightness of the planet are indicated by the relative sizes of the circles marking Mercury's position at five-day intervals. During its period of visibility its magnitude fades from −1.2 to 0.0; thus Mercury is at its brightest before it reaches greatest eastern elongation (21°) on May 4.

VENUS, continues to be visible for a short while in the evenings, after sunset, low in the western sky. It has a magnitude of −3.9.

MARS, magnitude +1.6, is visible for a short while low above the western horizon in the evenings. In early April, Mars moves eastwards from Aries into Taurus, passing between the two open clusters, the Hyades and the Pleiades. On April 19, Mars passes 6° north of Aldebaran.

JUPITER, magnitude −2.1, continues to be visible as a conspicuous evening object in the south-western sky. Observers in the latitudes of the British Isles will be able to see the planet until midnight.

SATURN, magnitude +0.1, continues to be visible as an evening object in the western sky. It is getting closer to the Sun and by the end of April

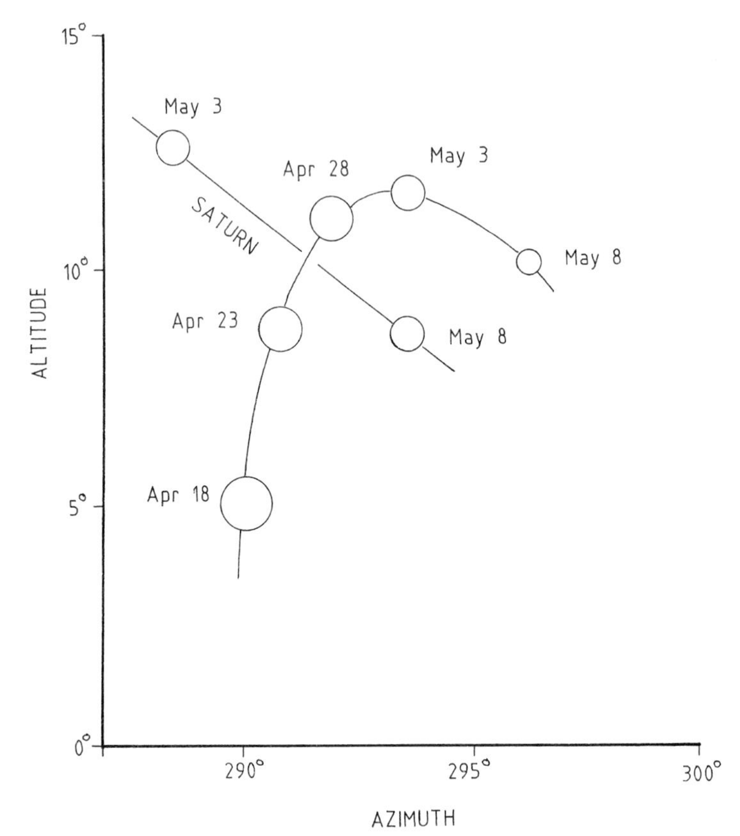

Figure 5. Evening apparition of Mercury, from latitude 52°N.

observers in the latitudes of the British Isles will not be able to see it for more than two hours after sunset. On April 16 these same observers, using a telescope, will be able to see a daylight occultation of Saturn by the Moon shortly before sunset.

The Largest and smallest constellations. During April evenings, the large, faint constellation of Hydra, the Watersnake, sprawls across the southern sky, beginning not far from Procyon in Canis Minor and ending below Spica in Virgo. It has only one bright star, Alphard or Alpha Hydræ (often called 'the Solitary One' because there are no conspicuous stars anywhere near it), which is of the second magnitude. The rest of Hydra consists of a straggling line of faint stars below Leo and Virgo. Extensive though they are, such large and faint constellations attract little notice in comparison with much smaller constellations that have brighter stars.

There is little order and method in the division of the sky into its various constellations; indeed, it has been claimed that the system must have been devised so as to cause as much confusion and inconvenience as possible! Of course, some of the constellations were formed in ancient times – Ptolemy, who died in or about AD 180, listed a total of 48 – while others have been added much more recently. Argo Navis, which used to hold pride of place with regard to area, was unceremoniously chopped up, and now, in area, Hydra is the largest of the constellations. These areas are not difficult to calculate nowadays, because the boundaries of all the constellations are straight lines, or, to be more precise, arcs of great circles on the celestial sphere. Before 1930 the boundaries were left to the fancy of the artists who designed the old pictorial star atlases, and there was little general agreement as to how the boundaries should be drawn to include the outstretched arm of a figure, or the tail of an animal. In 1930, under the direction of Eugene Delporte of the Brussels Observatory, the boundaries were fixed as regular lines, and now it is definitely possible to assign any particular star to one constellation only. Of the 88 constellations, Hydra is the largest, followed by Virgo, Ursa Major and Cetus; the smallest are Sagitta and Equuleus in the northern hemisphere, and the famous Southern Cross, Crux Australis, near the South Pole.

The largest constellations are as follows:

	Name	Area, square degrees
1.	Hydra	1303
2.	Virgo	1294
3.	Ursa Major	1280
4.	Cetus	1231
5.	Hercules	1225
6.	Eridanus	1138
7.	Pegasus	1121
8.	Draco	1083
9.	Centaurus	1060
10.	Aquarius	980

At the other end of the scale come:

	Name	Area, square degrees
1.	Crux Australis	68
2.	Equuleus	72
3.	Sagitta	80
4.	Circinus	93

Remarkably, Orion, the Hunter, has an area of just 594 square degrees, and comes a mere 25th in order of size, despite its exceptional prominence and its collection of bright stars.

Argo Navis. So far as observers in Europe and the northern part of the United States are concerned, early evenings in April give the best opportunity of seeing a few stars of the huge constellation of Argo, the Ship – in mythology, the vessel that carried Jason and his companions, the Argonauts, in their successful quest to remove the golden fleece of the ram (Aries) from its sacred grove, where it was guarded by a dragon.

Star maps used to include Jason's ship, the great Argo Navis, but it was so large and unwieldy that in 1932 the International Astronomical Union divided it into three separate parts, Puppis (the Poop), Carina (the Keel) and Vela (the Sails). Canopus, the second brightest star in the sky, which used to be known officially as Alpha Argûs, is now known as Alpha Carinæ. Unfortunately, Carina, the brightest part of the now-dismembered constellation of Argo is far south in the sky; but from the latitudes of London or New York, some of the stars of Puppis may be seen well above the horizon, lying to the east of Canis Major, the larger of Orion's two hunting dogs. The most northerly of the brighter stars of Puppis is Rho or Turais, which is variable over a very small range from magnitude 2.72 to 2.87. From southern Britain, when due south, Rho Puppis is 15 degrees above the horizon.

May

New Moon: May 12 *Full Moon*: May 26

MERCURY is at greatest eastern elongation (21°) on May 4 and therefore continues to be visible as an evening object for observers in tropical and northern latitudes for the first week of the month, low above the west-north-western horizon for a short while around the end of evening civil twilight. During this period its magnitude fades from 0.0 to +1.0. Observers should refer to Figure 5 (given with the notes for April). For the rest of the month it is too close to the Sun for observation, passing through inferior conjunction on May 27. It is not suitably placed for observation by those in more southerly latitudes.

VENUS is visible as a brilliant object in the western sky after sunset, magnitude −3.9. During May, Venus is moving steadily eastwards at a rate of more than 1° a day. In doing so it passes Saturn on May 7 and Mars on May 10. On the latter occasion Venus passes just north of Mars and thus provides a good opportunity to assist in locating the fainter planet. On May 4, Venus passes 6° north of Aldebaran. Observers should note that around May 14 the thin crescent Moon is visible in the western evening sky, together with the planets Venus, Mars, Jupiter and Saturn, though the latter object may be difficult to locate.

MARS continues to be visible as an evening object low above the west-north-western horizon for a short while after dusk. Its magnitude has now faded to +1.7. Mars and Saturn are quite close together at the beginning of the month and on May 4, Mars passes 2° north of Saturn: the latter planet is several times brighter and can therefore be used as a guide to locating Mars. Mars is reaching the end of its evening apparition for observers in the latitudes of the British Isles, and they will lose it in the evening twilight shortly before the end of the month.

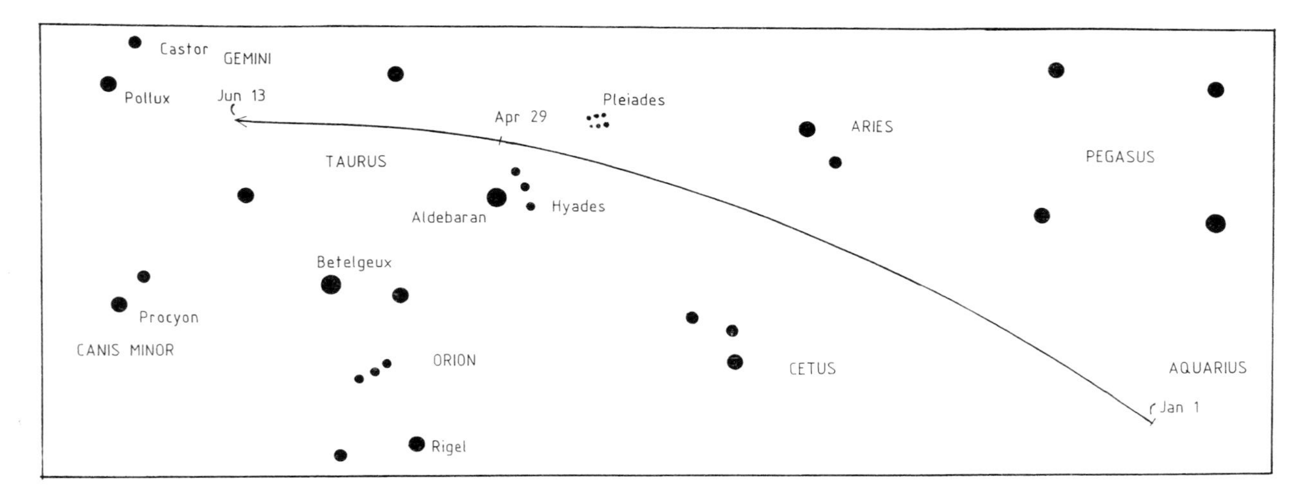

Figure 6. The path of Mars from January to June.

JUPITER continues to be visible as an evening object, in the south-western sky, though by the end of the month it is only visible for about two hours after sunset. Jupiter's magnitude is −1.9.

SATURN, magnitude +0.1, is now reaching the end of its evening apparition as it becomes increasingly difficult to detect low in the western sky in the gathering evening twilight. It is unlikely to be visible after the middle of the month.

Boötes and the Northern Crown. In late evenings in May the constellation of Boötes (the Herdsman) is high up in the south. Boötes is easily recognized because it is dominated by the brilliant orange star Arcturus, which may be found by following round the curve of the handle of the Plough (or the tail of the Great Bear), which at this time of the year is high in the sky, almost overhead. Boötes contains some other fairly bright stars, including Epsilon (magnitude 2.4), Eta (2.7) and Gamma (3.0). Both Arcturus and Eta Boötis are relatively near neighbours of our Sun, lying at distances of 36 and 32 light-years, respectively.

Slightly to the east of Boötes is the little semi-circlet of stars known as Corona Borealis, the Northern Crown. This distinctive little constellation is meant to represent the crown given by Bacchus to Ariadne, daughter of King Minos of Crete. This has always been a favourite group with stargazers, and its stars have more pet names and classical allusions than most other patterns. Its brightest star, Alpha or Alphekka, is also known as Gemma and Margarita Coronae. It is 130 times more luminous than our Sun, has an apparent magnitude of 2.2, and lies at a distance of 78 light-years. The second brightest star, Beta or Nusakan, at 28 times the Sun's luminosity, is rather closer at 59 light-years; its magnitude is 3.7.

In the bowl of the Northern Crown is the star R Coronae, which is usually just visible to the naked eye, but which at irregular intervals drops rapidly to a minimum so faint that moderate-sized telescopes are needed to see it, and then takes from several weeks to several months to recover. R Coronae variables are deficient in hydrogen but rich in carbon; their unpredictable fadings are due to clouds of 'soot' accumulating in the star's atmosphere.

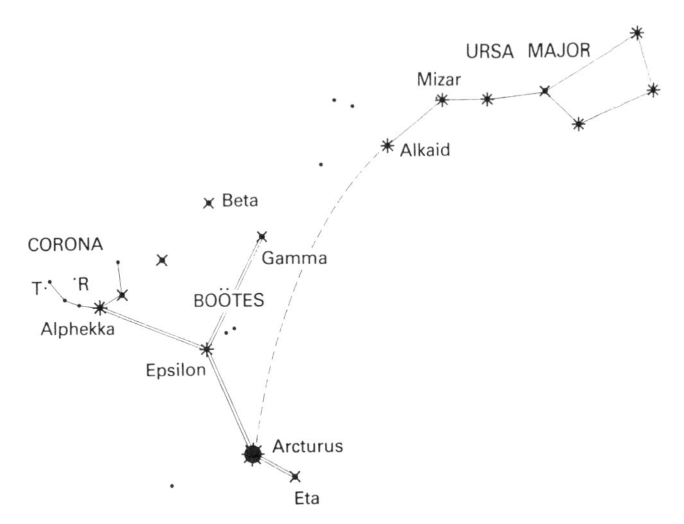

Figure 7. Boötes and Corona Borealis.

The Celestial Crow. From northern temperate latitudes, Corvus, the Crow, may be seen low in the south during May evenings. It contains no star brighter than magnitude 2.5, but its four leaders make up a well-marked quadrilateral, and the constellation can be quite prominent, particularly as it lies in a barren region.

To find Corvus, continue the long curve from the Great Bear's tail through Arcturus and Spica, and the distinctive shape of Corvus will be found easily enough. The brightest star, Gamma, has the proper name of Minkar; it is of spectrum B8, so that it is bluish-white, and is 250 times as luminous as our Sun. Its distance from us is about 185 light-years.

In 1603 Johann Bayer allotted Greek letters to the stars in each constellation, making the brightest star Alpha, the second brightest Beta, and so on. In some cases, however, the order of brightness has not been followed, and Corvus is a good example of this. Following Gamma in order of brilliancy are Beta (magnitude 2.7), Delta (2.9) and Epsilon (3.0); Alpha Corvi is considerably fainter at magnitude 4.0.

Below Corvus is part of the immensely large, very dim constellation of Hydra (the Watersnake). Lower still, and further east, observers in the latitude of London or New York may be able to make out a few stars of Centaurus (the Centaur), but unfortunately the greater part of this splendid constellation is too far south to rise over Europe or most of the United States.

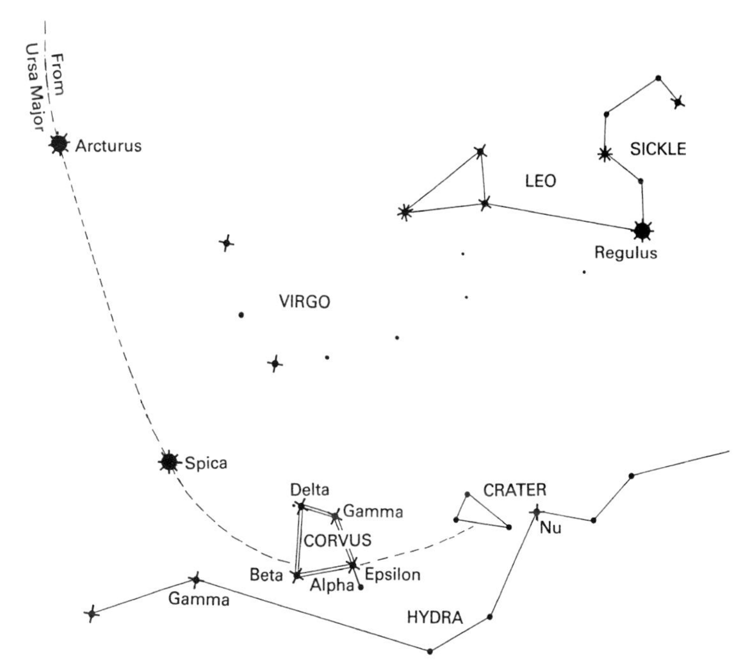

Figure 8. How to find Corvus, the Crow, using Arcturus and Spica.

June

New Moon: June 10 *Full Moon*: June 24

Solstice: June 21

MERCURY attains its greatest western elongation (23°) on June 21. For observers in the latitudes of the British Isles the long duration of twilight makes observation impossible, but nearer the equator and in the southern hemisphere Mercury can be seen as a morning object, low above the eastern horizon at the beginning of morning civil twilight, after the middle of the month. During this period the magnitude of Mercury brightens from +1.0 to −0.3.

VENUS, is a brilliant object in the western sky in the evenings, magnitude −4.0. Observers in the latitudes of the British Isles will notice that although it is continuing to increase its eastern elongation from the Sun, the time available for observation actually decreases slightly during the month. This is due to two reasons: the Sun is setting later at the end of the month than it is at the beginning, by about 15 minutes, and secondly, the planet is moving southwards in declination. On the evening of June 3, Venus passes 1.6° north of Jupiter. Six days later it passes 5° south of Pollux.

MARS, already lost to view by observers in the latitudes of the British Isles, can only be seen by observers further south for the first fortnight of the month, low above the west-north-western horizon in the early evening before twilight inhibits observation.

JUPITER, magnitude −1.9, is now coming towards the end of its period of visibility. It is visible only low in the western sky for a short while after sunset. Because of the long duration of twilight, observers in the latitudes of the British Isles are unlikely to be able to see the planet after the middle of the month.

SATURN remains too close to the Sun for observation throughout the month, as it passes through conjunction on June 9. It is of some interest to note that this month, as seen from the Earth, the rings are only 0.2° short of their maximum opening, which will occur next April.

PLUTO is at opposition on June 7, in the constellation of Ophiuchus. It is only visible with a moderate-sized telescope since its magnitude is +14. At opposition the planet is 4,414 million kilometres (2,743 million miles) from the Earth.

The Midnight Sun. In northern temperate latitudes, the Sun reaches its highest point in the sky at the solstice on June 21, when it will be 23.5° north of the equator. It then rises at its maximum bearing north of due east, and sets north of west, taking more than 12 hours for its passage across the sky, so that we have long days and short nights at this time of year. From the astronomer's point of view, there is no night at all in these latitudes, for by definition astronomical twilight does not end until the Sun is 18° below the horizon. In the summer months in this country, the Sun cannot sink to such an angle, even below the northern horizon, so that it never becomes completely dark. Thus in London at the solstice, the Sun is only 15° below the northern horizon at midnight, while in Aberdeen the depression at midnight is only about 9°. The following table shows the dates between which there is no true night at different latitudes:

Latitude	**Date**
51° (Southern England)	May 26 – July 18
53° (Midlands)	May 16 – July 28
55° (Northern England)	May 8 – August 5
57° (Scotland)	May 1 – August 12

Still further north, in latitudes greater than 66.5°, there will always be at least one day in the year when the Sun does not set, but merely circles the sky during the whole 24 hours. This is the so-called 'midnight sun', and a glance at the atlas will show that there are many countries that might claim to be the 'Land of the Midnight Sun' in the northern hemisphere. These include the most northerly parts of Norway, Sweden, Finland, Russia, Alaska and Canada, and most of Greenland. In southern latitudes, the midnight Sun will be visible only from Antarctica.

Libra. The constellation of Libra (the Scales), which is well placed during late evenings in June, is one of the zodiacal constellations but is otherwise undistinguished. Libra originally formed the right-hand part of Scorpius, and was known as Chelae Scorpionis (the Scorpion's Claws). In the first century BC it became known as Libra, the Scales or Balance, presumably because at that time the Sun was in this part of the sky at the autumn equinox, when day and night are equal. The equality thus suggested the idea of the scales, although they are not exactly balanced, as drawn in the old star atlases.

The three main stars of Libra form a triangle with the point upwards. The Arabic names of two of these stars remind us of the original constellation. The right-hand star of the triangle, Alpha Librae (magnitude 2.7), bears the name Zuben el Genubi, meaning the Southern Claw. Because it lies almost exactly on the ecliptic, it was used in ancient times as one of the marks of progress of the Moon and planets round the sky. More recently it was used as one of the 57 navigational stars. Zuben el Genubi is a double star, but although the two stars are widely separated, they are of unequal brightness (magnitudes 2.8 and 5.2), and require a pair of field-glasses to distinguish them.

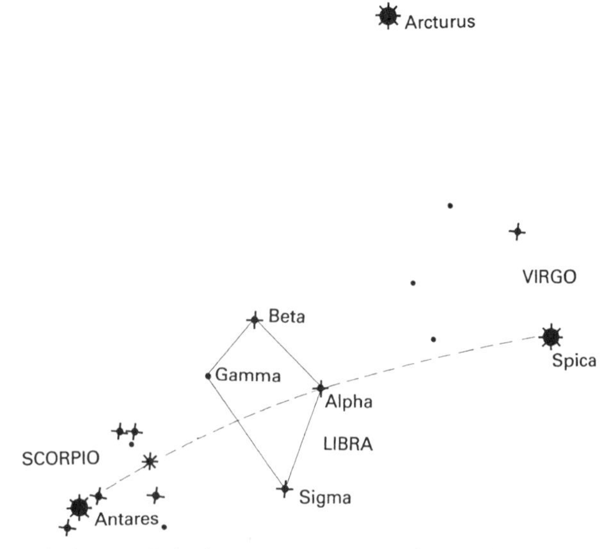

Figure 9. The location of Libra between Spica in Virgo and Antares in Scorpius.

Beta Librae (magnitude 2.6), at the top of the triangle, has the Arabic name Zuben el Chemale (the Northern Claw), and it is said to be the only bright single star which appears green in colour, but most observers will just regard it as white. The third star, Gamma Librae, at the left of the triangle, is sometimes given the name Zuben el Hakrabi; it is significantly fainter than the other two at magnitude 3.9. Sigma Librae (magnitude 3.3), which lies below the triangle formed by the other three, was formerly included in Scorpius, as Gamma Scorpii.

July

New Moon: July 10 *Full Moon*: July 24

EARTH is at aphelion (farthest from the Sun) on July 6 at a distance of 152 million kilometres (94.5 million miles).

MERCURY, for observers in equatorial and southern latitudes, continues to be visible as a morning object, low above the eastern horizon for the first week of the month. During this period its magnitude brightens from −0.3 to −0.9. Thereafter it is too close to the Sun for observation, as it passes through superior conjunction on July 21.

VENUS, magnitude −4.1, continues to be visible as a brilliant object in the western sky after sunset. Venus passes 1° north of Regulus on July 10.

MARS continues to be unsuitably placed for observation.

JUPITER is unsuitably placed for observation, as it passes through conjunction on July 20.

SATURN, magnitude +0.1, is slowly emerging from the morning twilight, becoming visible low above the eastern horizon for a short while before the morning twilight inhibits observation. For observers in the latitudes of the British Isles the long duration of twilight at this time of year means that they are very unlikely to see it before the middle of the month.

Seasonal variations. The Earth's orbit round the Sun is not perfectly circular; it is an ellipse of very slight eccentricity, with the Sun located at one of the two foci of the ellipse. Consequently, the orbit is not centred on the Sun, so that the Earth's distance from it varies during the course of the year. In early July the Earth is at aphelion (farthest from the Sun), and the distance is 152.1 million kilometres (94.5 million miles). At perihelion (closest to the Sun) in early January, the distance is reduced to 147.1 million kilometres (91.4 million miles). So the Earth receives just 6.9 per cent more solar radiation at perihelion than it does at aphelion.

It may seem curious that aphelion occurs in the northern latitudes' summer, but in point of fact the seasons have very little to do with the Earth's changing distance from the Sun. They are caused by the fact that the axis of rotation is tilted at about 23.5°; in northern winter, the North Pole is tilted away from the Sun, whereas in northern summer the North Pole is tilted toward the Sun, thus receiving the solar rays much more directly. The slight effects of the varying distance are masked by geographical peculiarities of the Earth, inasmuch as there is a great deal more ocean in the southern hemisphere.

Conditions on the planet Mars are roughly analogous. The axial tilt is almost 24°, much the same as that of the Earth, and so the seasons are of the same general type, though they are much longer (the Martian 'year' is 687 Earth-days, though the axial rotation is a little over half an hour longer than ours). There is, however, one important difference. The orbit of Mars is considerably more eccentric than that of the Earth, so that its varying distance plays a fairly major rôle. When Mars is at aphelion, its distance from the Sun is 249.1 million kilometres (154.8 million miles), whereas at perihelion, the distance is 206.7 million kilometres (128.4 million miles). So Mars receives 45 per cent more solar radiation at perihelion than at aphelion. Consequently, the southern winters, when Mars is near aphelion, are significantly longer and colder than those of the northern hemisphere. This is shown by the behaviour of the polar caps. The southern cap may become larger than the maximum extent of its northern counterpart – but at Martian midsummer it may disappear completely, whereas the northern cap never does so.

Lyra. Vega, or Alpha Lyrae, occupies the overhead position late on July evenings, and its brilliant bluish hue makes it stand out at once. At a distance of just 25 light-years,Vega is a comparatively near neighbour of our Sun on the cosmic scale. The fifth brightest star in the night sky, Vega is the only really bright star in Lyra (the Lyre or Harp), but the constellation, though small, contains more than its fair share of interesting subjects.

Particularly noteworthy is the famous 'double-double' or quadruple star Epsilon Lyrae, close to Vega. Ordinary-sighted people can see the two main components (magnitudes 4.7 and 5.1, separation 208 arc seconds) with the naked eye, and each of these is itself double (separations 2.6 and 2.3 arc seconds, respectively). A 3-inch refractor will show all four components. Zeta Lyrae is also a wide, easy telescope double (magnitudes 4.4 and 5.7, separation 44 arc seconds).

Beta Lyrae, sometimes still known by its old proper name of Sheliak, is a remarkable system. It is an eclipsing binary, but quite different from Algol, the Demon Star, in Perseus. Beta Lyrae's variations are continuous, and there are alternate deep and shallow minima, one of magnitude 3.8 and the other of magnitude 4.3; the period is 12.94 days. Maximum is only 3.3, so that Beta is never a prominent naked-eye object; at its brightest it is about equal to Megrez, the faintest of the seven main stars in the Great Bear. The two components are so close together that they almost touch, and so cannot be seen separately. Each must be stretched out into the shape of an egg, while there is an extensive gaseous envelope. Beta Lyrae has received an immense amount of attention from astrophysicists, since it is by far the brightest member of its class. The variations may be followed with the naked eye; the nearby star Gamma Lyrae (3.2) is a convenient comparison.

Between Beta and Gamma lies the celebrated Ring Nebula, Messier 57, one of the brightest planetary nebulae. It is below naked-eye visibility, but a 3-inch refractor will show it distinctly, and it is easy to find in view of its position midway between Beta and Gamma. The central star is a very elusive object, and requires large apertures. R Lyrae, forming an obtuse triangle with Vega and Delta Cygni in the 'Northern Cross' is an M-type semi-regular variable; the range is from 3.9–5.0, and there is an approximate period of 46 days.

August

New Moon: August 8 *Full Moon*: August 22

MERCURY reaches its greatest eastern elongation (27°) on September 1 and is visible throughout August to observers in equatorial and southern latitudes for whom it is the most suitable evening apparition of the year. Figure 10 shows, for observers in latitude 35° S, the changes in azimuth (true bearing from the north through east, south and west) and altitude of Mercury on successive evenings when the Sun is 6° below the horizon. This condition is known as the end of evening civil twilight, and in this latitude and at this time of year occurs about 30 minutes after sunset. The changes in the brightness of the planet are indicated by the relative sizes of the circles marking Mercury's position

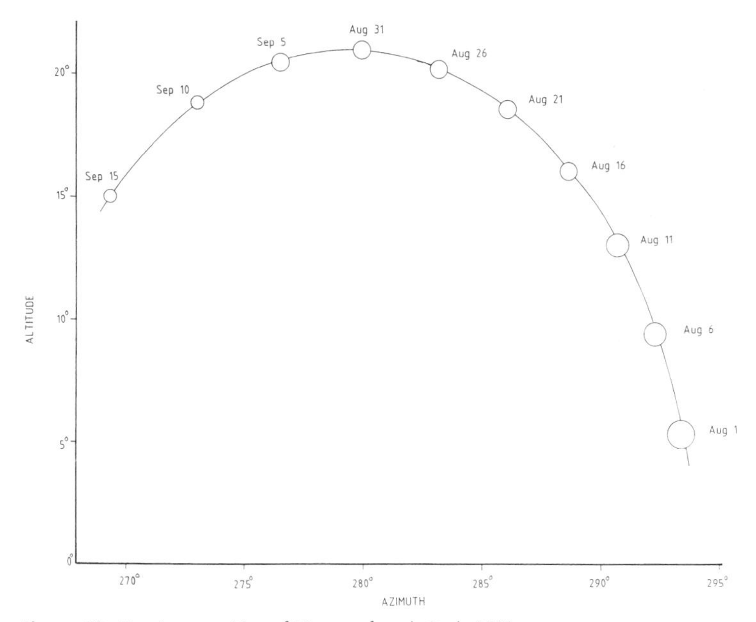

Figure 10. Evening apparition of Mercury, from latitude 35°S.

at five-day intervals. During its period of observation Mercury's magnitude fades from −0.9 to +0.2. It will be noticed that Mercury is at its brightest before it reaches greatest eastern elongation. Mercury is not visible to observers in the latitudes of the British Isles during August.

VENUS reaches greatest eastern elongation (46°) on August 22, with a magnitude of −4.3. Thus it is a magnificent object in the western sky after sunset. The visibility conditions are quite different in the two hemispheres: whereas observers in latitude 52° N will only be able to see the planet for about 50 minutes after sunset, even at the beginning of the month, falling to less than 20 minutes by the end of the month, those in latitude 35° S should be able to see it for around three hours after sunset.

MARS passes through conjunction on August 10, and thus remains unsuitably placed for observation.

JUPITER, magnitude −1.8, becomes visible in the early morning skies during August, low above the eastern horizon for a short while before sunrise. It becomes visible to observers in the British Isles after the first ten days of the month and almost a week later to observers further south. Jupiter is in the constellation of Cancer.

SATURN, magnitude +0.1, is a morning object, visible low above the eastern horizon in the early hours.

URANUS is at opposition on August 20. It is barely visible to the naked eye as its magnitude is +5.7, but it is readily located with modest optical aid. At opposition Uranus is 2,841 million kilometres (1,765 million miles) from the Earth. Uranus is on the Capricornus–Aquarius border.

NEPTUNE is at opposition on August 2, in the constellation of Capricornus. It is not visible to the naked eye since its magnitude is +7.8. At opposition the planet is 4,349 million kilometres (2,702 million miles) from the Earth.

'The Georgian Planet'. Uranus, at opposition during this month, was the first major planet to be discovered in modern times. It was found in

1781 by William Herschel, then a relatively unknown amateur, but afterwards perhaps the greatest of all astronomical observers.

Five planets (Mercury, Venus, Mars, Jupiter and Saturn) had been known since ancient times. With the Sun and Moon, the total number of bodies in the Solar System was therefore seven; and since seven was the 'mystical' number, it was tacitly assumed that no more planets could exist. Certainly, Herschel was not deliberately looking for one. He was conducting a systematic review of the sky, using reflecting telescopes of his own construction, when he noticed an object in the constellation Gemini that showed a definite disk, and could not therefore be a star. Naturally, he lost no time in reporting the discovery, but he did not realize that he had found a new planet; his original paper was entitled 'An Account of a Comet'. Not until the movements of the object had been studied, and the orbit worked out, did the full significance of the discovery become known.

Herschel suggested calling his planet the 'Georgium Sidus', in honour of King George III of England and Hanover. (Herschel was Hanoverian by birth, though he had come to England while still a young man and spent the rest of his life there.) The name was not well received, and for a time the planet was called Herschel; but eventually the name Uranus, proposed by the German astronomer, Johann Bode, came into general use.

Uranus is a giant planet, with an equatorial diameter of 51,118 kilometres (31,764 miles), and a mass 14.6 times that of the Earth. It takes 84 years to complete one orbit of the Sun. The most remarkable feature is the tilt of the planet's axis of rotation, which amounts to almost 98° – more than a right angle – so that technically Uranus rotates in a retrograde or 'wrong-way' direction. Sometimes the pole appears in the middle of the disk, as viewed from Earth, while at other times the equator is presented.

Uranus is not a spectacular object, but it is interesting to find. The planet can just be seen with the naked eye under good conditions. Any small telescope will show its disk, but large instruments are needed to bring out any surface detail.

Mariner 2 and the First Fly-by of Venus. This month we mark the 40th anniversary of the launch of the US space probe Mariner 2, which made the first successful fly-by of the cloud-covered planet Venus in December 1962. Mariner 2 blasted off on August 27, 1962, and while en route to Venus it detected a continuous stream of electrified particles being sent out all the time by the Sun. It also observed fast and slow streams which repeated at 27-day intervals, suggesting that the source of this 'solar wind' rotated with the Sun. On arrival at Venus, which it passed at a distance of 34,800 kilometres (21,620 miles) on December 14, 1962, Mariner 2 revolutionized many of our ideas about the planet. Prior to its fly-by, there was a theory, proposed by Fred Whipple and Donald Menzel, that Venus was mainly covered in oceans, and the clouds were composed chiefly of water droplets. Mariner 2 killed off this theory at once, confirming the high surface temperature and long axial rotation period of 243 days. There was also no detectable magnetic field. Mariner 2 showed that Venus and Earth were non-identical twins, and its success marked the beginning of the era of direct planetary exploration by spacecraft. Contact with Mariner 2 was lost on January 4, 1963.

September

New Moon: September 7 *Full Moon*: September 21

Equinox: September 23

MERCURY reaches greatest eastern elongation (27°) on September 1 and passes through inferior conjunction on September 27, but continues to remain unsuitably placed for observation by those in the latitudes of the British Isles. Further south the planet continues to be visible as an evening object for the first fortnight of the month and reference should be made to Figure 10 (given with the notes for August). During this period its magnitude fades from +0.3 to +1.0.

VENUS, magnitude −4.6, continues to be visible as a magnificent object in the western sky, reaching its greatest brilliancy on September 26. Sadly for observers in the British Isles, Venus's increasing southern declination during September means that the planet will only be visible for a short while after sunset early in the month and will be too close to the Sun for observation after the middle of the month.

MARS, magnitude +1.8, remains too close to the Sun for observation by those in southern latitudes. However, observers in northern temperate latitudes may be able with difficulty to detect the planet as a morning object, low above the eastern horizon for a very short while before twilight inhibits observation. Mars is in the constellation of Leo.

JUPITER continues to be visible as a conspicuous morning object in the south-eastern sky, magnitude −1.9. Jupiter is in the constellation of Cancer.

SATURN continues to be visible as a morning object, and can now be seen in the eastern sky well before midnight. Its magnitude is +0.1. Saturn is in the constellation of Taurus and its position among the stars is shown in Figure 4 (given with the notes for March).

The Harvest Moon. In northern temperate latitudes, the Full Moon of September is generally referred to as the Harvest Moon, although the name is more correctly given to the Full Moon which occurs nearest to the autumnal equinox, and which may, therefore, come in early October. This September, Full Moon occurs on September 21, just two days before the autumnal equinox. For several nights in succession at this time of year the Moon rises only a few minutes later each evening, the delay from night to night being noticeably less than usual. The phenomenon of the Harvest Moon is repeated to a lesser extent at the following Full Moon (called the 'Hunter's Moon') as a result of the ecliptic being at its lowest in the sky on autumn evenings and being inclined at a very shallow angle to the horizon.

The average daily motion of the Moon is about 13°, and if the Moon's path were perpendicular to the horizon, moonrise would be delayed from night to night by the time taken by the Earth to turn through 13°, that is, by 52 minutes. If the angle that the Moon's path makes with the horizon is small, then the delay is much reduced, and under favourable conditions may be only about 9 minutes. Since the Moon travels right round the sky once a month, the effect occurs every month, but our attention is drawn to it most effectively at the time of Full Moon, which rises at the convenient time of sunset. It is thus in the autumn months that we are aware that there is more moonlight than usual, a fact that was of benefit to farmers in the days before the advent of tractors with headlamps. It should be noted that there is no Harvest Moon in the tropics, while the phenomenon occurs in March in the southern hemisphere.

Delphinus. One of the most prominent of the small constellations, and well placed during September evenings, is Delphinus, the Dolphin. It lies roughly between Altair and the Square of Pegasus. It has no star brighter than Beta, of magnitude 3.5, but it is a beautifully compact little group, and is quite unmistakable. According to an old legend, it represents the dolphin that rescued the famous singer Arion, who had been thrown into the sea by a party of sailors anxious to lay their hands upon his riches. The dolphin brought Arion safely to shore, and was rewarded by being placed in the sky!

The stars Alpha and Beta Delphini have the curious names Svalocin and Rotanev, respectively. These were allotted to the stars by one Nicolaus Venator, for reasons that should be obvious! Alpha is 60 times more luminous than our Sun, and lies at a distance of 170 light-years, while Beta is 46 times as luminous and is 108 light-years distant.

One of the stars in Delphinus, Gamma, is double; its components are of magnitudes 4.5 and 5.5, and 9.2 seconds of arc apart. The primary is yellow, the companion greenish. Gamma Delphini is a beautiful object in a moderate telescope.

Delphinus is not the only small constellation in the region, but the others – Equuleus (the Little Horse), Sagitta (the Arrow) and Vulpecula (the Fox) are much less conspicuous. Vulpecula does however include the Dumb-bell Nebula (Messier 27), which is a fine example of a planetary nebula. It is too faint to be well seen in small telescopes, but shows considerable detail when photographed with large instruments.

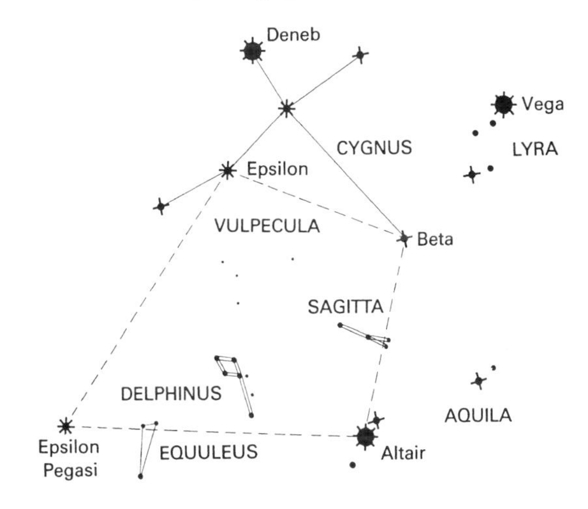

Figure 11. How to find Delphinus using Altair in Aquila, Cygnus, and the star Epsilon Pegasi.

October

New Moon: October 6 *Full Moon*: October 21

Summer Time in the United Kingdom ends on October 27.

MERCURY reaches greatest western elongation (18°) on October 13 and is visible as a morning object from between about October 6 to October 25. For observers in northern and equatorial latitudes this will be the most favourable morning apparition of the year. Figure 12 shows, for observers in latitude 52° N, the changes in azimuth (true bearing from the north through east, south and west) and altitude of Mercury on successive mornings when the Sun is 6° below the horizon. This condition is known as the beginning of morning civil twilight and in this latitude, and at this time of year occurs about 35 minutes before

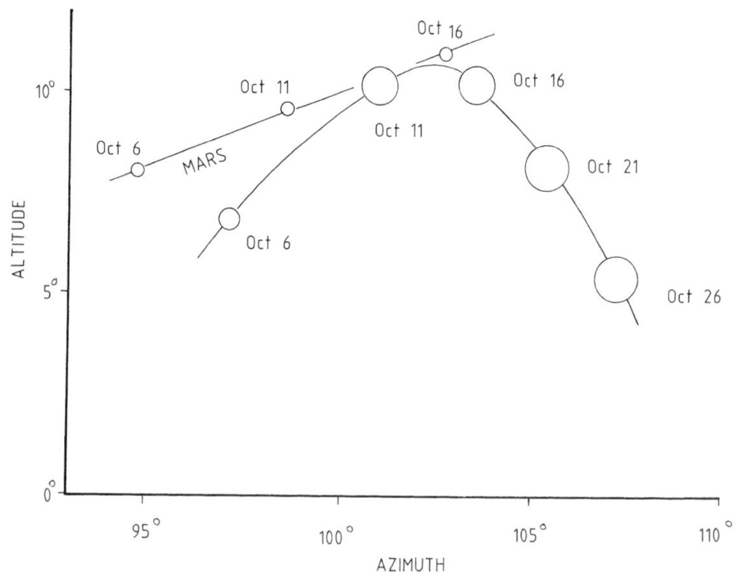

Figure 12. Morning apparition of Mercury, from latitude 52°N.

sunrise. The changes in the brightness of the planet are indicated by the relative sizes of the circles marking Mercury's position at five-day intervals. During this period its magnitude brightens from +1.0 to −1.0; thus Mercury is at its brightest after it reaches greatest western elongation. For some days around October 10, Mercury will only be a few degrees east of Mars.

VENUS is no longer visible to observers in the latitudes of the British Isles. However, for those further south, Venus continues to be visible as a magnificent object, magnitude −4.4, dominating the western sky after sunset, though just before the end of the month it becomes lost in the sunset glow.

MARS is still only visible to observers in the northern hemisphere, where it is visible in the pre-dawn sky, low above the east-south-eastern horizon, magnitude +1.8. Mars moves eastwards from Leo into Virgo early in October. The path of Mars among the stars for the rest of the year is shown in Figure 14 (given with the notes for November).

JUPITER, magnitude −2.0, continues to be visible as a conspicuous object in the south-eastern sky in the mornings until shortly before sunrise.

SATURN, magnitude −0.1, is still visible as a morning object. On October 11 it reaches its first stationary point, in the constellation of Taurus, some 15° north of Betelgeuse.

The Great Square of Pegasus. In northern temperate latitudes, the Square of Pegasus dominates the southern part of the evening sky during the autumn months of October and November. On star maps, Pegasus (the Flying Horse) looks quite conspicuous, since its four main stars make up an obvious square, and you might think that it would stand out easily in the night sky. Certainly, once you have found Pegasus, it will be easily located again, but to begin with you must look for it carefully. This is partly because its stars are not remarkably brilliant – all four are between the 2nd and 3rd magnitudes – and partly because the Square is very large. The 'W' of Cassiopeia, which lies almost overhead during autumn evenings, makes a useful pointer: an imaginary line passing through its stars Gamma and Alpha will lead you straight to the Great Square.

The upper left-hand star of the Square used to be known as Delta Pegasi, but now that it has been officially – and frankly, illogically – transferred to the neighbouring constellation of Andromeda, it is called Alpha Andromedae; its proper name, still often used, is Alpheratz. The upper right-hand star of the Square, Beta Pegasi or Scheat, is a red giant. Like many of its kind, it is variable; the magnitude range is from 2.3 to 2.8, and there is a very rough period of from 35 to 40 days. Alpha Pegasi or Markab (2.5) at the bottom right of the Square, and Gamma Pegasi or Algenib (2.9) at the bottom left, act as good comparison stars, and the light variations may be followed with the naked eye. The only other important star is Epsilon Pegasi or Enif (2.4), which lies well to the west of the Square, in a relatively isolated position roughly between Alpha Pegasi and Altair in Aquila, the Eagle.

It is interesting to count the number of faint stars that are visible inside the Square without optical aid. There are not a great many of them, though the area covered is considerable; anyone who counts a dozen will be doing rather well.

Well below the Square, fairly close to the horizon in Britain, is Fomalhaut in Piscis Austrinus (the Southern Fish), the southernmost of the first-magnitude stars visible from Europe or from most parts of the United States. It is never well seen from these locations, but fortunately a line extended downwards from Scheat through Markab points almost directly to it. The best time to look for Fomalhaut is around 2000 to 2100 hours in mid-October; by midnight it will be too low to be seen, although from the latitude of New York it is visible for longer.

The Milky Way. For observers in Europe and North America, October evenings present the Milky Way at its best. The glorious shining band stretches from one horizon to the other, and passes almost overhead in the neighbourhood of Cassiopeia. The dark 'rifts' in the Cygnus area are easily identified even with the naked eye.

There are many legends about the Milky Way, but it was not until the start of telescopic astronomy that its nature was realized. Galileo, soon after he made his first primitive 'optick tube', recorded that the Milky Way is made of countless faint stars, which seem to be so close together that they almost touch. It is now known that the crowding is merely a line of sight effect. The Galaxy is a flattened system, and when looking along the main plane we see many stars in almost the same direction. The dark 'rifts' that are visible here and there within the band of the Milky Way indicate not an absence of stars, but rather the presence of dark, obscuring material.

The distance of the Sun from the galactic centre is thought to be about 27,000 light-years. We cannot see the actual centre, which lies in the direction of the lovely star clouds seen in the constellation of Sagittarius; there is too much obscuring material in the way. Our knowledge of it is drawn from infra-red and radio observations, since the dust cannot block out radiations at these wavelengths.

The owner of a pair of binoculars can spend many enjoyable hours sweeping around the Milky Way area of the sky. Innumerable rich star-fields will come into view, together with clusters and groups of stars.

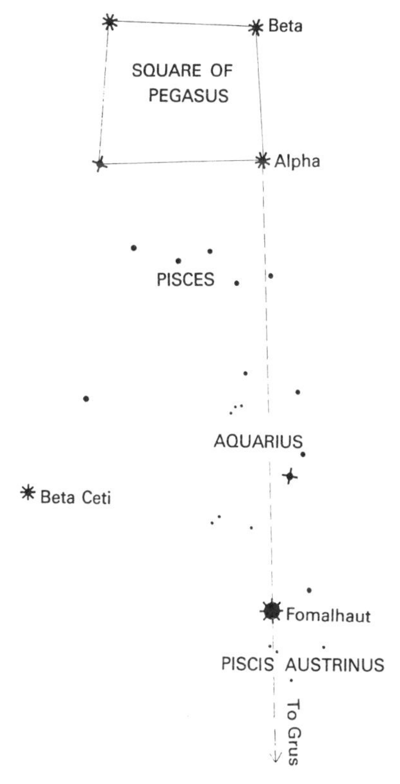

Figure 13. Finding the star Fomalhaut using the Great Square of Pegasus.

November

New Moon: November 4 *Full Moon*: November 20

MERCURY passes through superior conjunction on November 14, and is thus unsuitably placed for observation throughout the month.

VENUS, having just passed through inferior conjunction, becomes visible as a morning object, magnitude −4.4, though observers in the latitudes of the British Isles will have to wait until after the first week of the month to see it. It is then visible above the south-eastern horizon before dawn. Venus is now moving rapidly away from the Sun, and by the end of the month northern observers can detect the planet low above the horizon several hours before sunrise.

MARS, magnitude +1.8, is a morning object visible in the eastern sky well before dawn. Mars is moving steadily eastwards in the constellation of Virgo, passing 3° north of Spica on November 20, as can be seen from Figure 14, which shows the movement of Mars among the stars during the second part of the year.

JUPITER is still visible as a conspicuous morning object, magnitude −2.2, and, for observers in the latitudes of the British Isles, it is visible well before midnight. Jupiter moves eastwards from Cancer into Leo at the beginning of the month.

SATURN, magnitude −0.3, is now visible for the greater part of the night as it approaches opposition next month. The rings of Saturn present a beautiful spectacle to the observer with a small telescope. The Earth passed through the ring plane twice in 1995 and since then the rings have been slowly opening up: the diameter of the minor axis is now 20 arc seconds, marginally greater than the polar diameter of the planet itself. The rings will not be at their maximum opening until early next year.

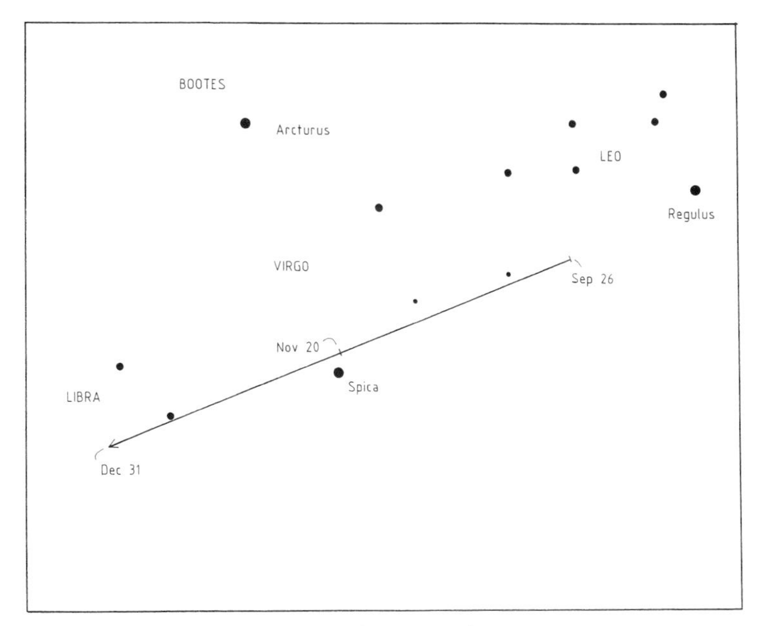

Figure 14. The path of Mars from September to December.

Capella and Vega. The two stars Capella and Vega are almost exactly equal in magnitude, and lie on opposite sides of the Pole Star, so that they may pass virtually overhead as seen from Britain and northern parts of the United States. Capella, the brightest star in Auriga, the Charioteer, occupies the place of honour during winter evenings; Vega in Lyra, the Lyre, during summer evenings.

During evenings in November the two stars are roughly equal in altitude, Vega sinking, Capella rising, so that it is interesting to compare them. Their colours are very different. Vega, an A-type star, is bluish; Capella is actually a very close binary star and both components are of spectral type G, so it is yellow. This means, of course, that Vega is the hotter of the two. Its surface temperature is more than 10,000° C, as against less than 5,000° C for the components of Capella. On the other hand, the components of Capella are 90 and 70 times the luminosity of our Sun, while Vega is only about 50 times brighter than the Sun. Their distances are, respectively, 42 light-years and 25 light-years.

Though Capella and the Sun are very similar in spectral type, there is a great difference between them. The two components of Capella are both about three times the mass of our Sun and much larger, with diameters of 18 million and 10 million kilometres (11 million and 6 million miles), while our Sun's diameter is just 1.4 million kilometres (865,300 miles). The most modern apparent magnitude measures give Vega +0.03 and Capella +0.08, so Vega is just the brighter of the two.

The only star in the northern hemisphere of the sky to surpass Capella and Vega in brilliance is Arcturus in Boötes (the Herdsman), with an apparent magnitude of −0.04. It is best seen on evenings during spring in the northern hemisphere.

Cassiopeia. During November evenings the prominent 'W' of stars marking Cassiopeia is almost overhead, and it is worth taking a good look at this distinctive constellation. It is circumpolar from Britain, so that it is always to be seen whenever the sky is sufficiently dark and clear, and its outline is unmistakable.

Of the five main stars making up the 'W' (or 'M' depending upon which way you look at it), Beta is of magnitude 2.3, Delta 2.7 and Epsilon 3.4. Of the remaining two, Alpha Cassiopeiae, sometimes still known by its old proper name of Shedir, is of about magnitude 2.2, but may be slightly variable. Gamma, the centre star of the 'W', is undoubtedly variable; it is a 'shell star' with an unusual spectrum. From all accounts, it remained at just below the second magnitude until 1936, when it abruptly brightened up until it surpassed the Pole Star, reaching magnitude 1.6. It then faded until it had dropped almost to the third magnitude. For some years now it has usually been rather similar in brightness to Alpha or Beta.

Needless to say, the stars in the 'W' are not really associated with each other. Gamma is the most distant, lying at 780 light-years from us, and Epsilon is rather closer at 520 light-years. Next comes Shedir at 120 light-years; its spectrum is of type K, and it is decidedly orange when observed with binoculars or through a telescope. Delta lies at 62 light-years, Beta at only 42.

The star Rho Cassiopeiae, situated not far from Beta, is a most unusual variable. Generally it is around magnitude 4.9, but it has been known to fade below 6. It seems to be extremely luminous, and very remote at 8,000 light-years. Its spectrum is variable, but generally near type F8.

The Milky Way flows through Cassiopeia, and the whole region is very rich, so that it is worth sweeping over with a low power and a wide field. In addition, there is a fine binary in Eta Cassiopeiae, and a fairly bright open cluster, Messier 52, which lies almost in a straight line with Shedir and Beta.

Mythologically, Cassiopeia was the proud queen (and wife of king Cepheus) who boasted that the beauty of her daughter, Andromeda, was greater than that of the sea-nymphs. Since the nymphs were the daughters of Neptune, god of the ocean and brother of Jupiter, this was a most unwise claim, and Neptune took the poorest possible view of it, sending a fearsome sea monster to ravage the queen's land. Complete disaster was averted only by the timely arrival of the hero Perseus, who obligingly turned the monster to stone.

December

New Moon: December 4 *Full Moon*: December 19

Solstice: December 22

MERCURY attains its greatest eastern elongation (20°) on December 26, and for observers in tropical and southern latitudes it is visible as an evening object low in the western sky at the end of evening civil twilight, for all but the first few days of the month. Observers in the latitudes of the British Isles and North America will find observation much more difficult, and it is only during the last few evenings of the month that they stand any chance of locating the planet, which is low in the south-western sky. During the month the magnitude of Mercury fades slowly from −0.6 to 0.0.

VENUS is a magnificent object in the early mornings, attaining its greatest brilliancy (magnitude −4.7) on December 7. For several hours before sunrise it completely dominates the south-eastern sky. At the beginning of the month Venus is about 2° east of Mars and at the end is still only about 5° east of the fainter planet. The minimum separation is 1.6° on the morning of the December 6. Have you ever seen Venus in daylight? Two good opportunities occur – on the mornings of December 1 and 30 when the old crescent Moon will be seen approaching the planet. These dates are favourable for locating Venus in daylight, shortly after sunrise, using the Moon as a guide; the observer should keep in the shadow.

MARS continues to be visible for several hours in the south-eastern sky in the early mornings, magnitude +1.6. Mars moves from Virgo into Libra during December, passing only 0.4° north of Alpha Librae on the 24th.

JUPITER remains a prominent morning object in the southern sky with a magnitude of −2.4. It is now visible in the east in the late evening. On December 4, Jupiter reaches its first stationary point, and with its retrograde motion returns from Leo back into Cancer before the end of the month.

SATURN, magnitude −0.5, reaches opposition on December 17 and is therefore visible throughout the hours of darkness. Saturn is moving slowly retrograde in the eastern part of the constellation of Taurus. At opposition Saturn is 1,205 million kilometres (749 million miles) from the Earth.

The Return of the Hunter. By December evenings, the brilliant winter constellations have again come into view. Of these, Orion, the Hunter, is pre-eminent: with its characteristic outline, prominent stars and surrounding retinue, it dominates the southern aspect of the sky.

Betelgeux, the red M-type giant in the upper left-hand part of the constellation, is a semi-regular variable which varies between magnitudes 0.1 and 0.9. Its distance is 310 light-years. Betelgeux – whose name can also be spelled Betelgeuse – is different from the other main stars of Orion, which are generally hot, white and of early spectral type (O or B). Rigel, in the Hunter's foot, which is of spectral type B8, is one such star. Rigel is actually variable over a very small range, but it has a mean apparent magnitude of +0.12, so it is less than one-tenth of a magnitude fainter than Capella and Vega, but when seen from Europe it does not appear their equal, simply because it never rises particularly high. Yet Rigel is vastly more luminous: according to recent estimates it is 60,000 times as luminous as the Sun, and lies at a distance of some 910 light-years.

Luminous though Rigel is, some of the other stars of Orion are

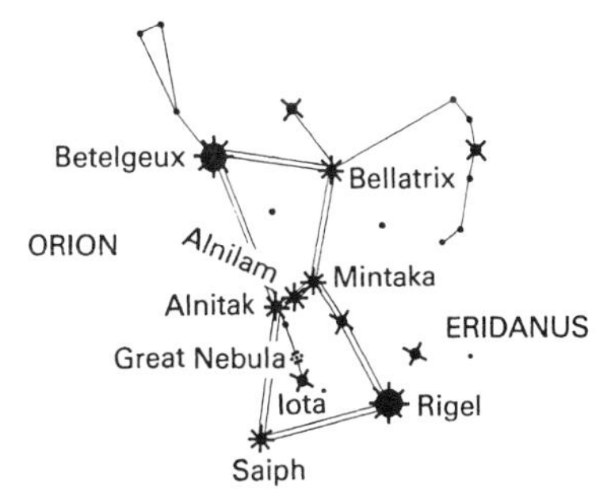

Figure 15. The principle named stars of Orion, the Hunter.

comparable. Kappa, or Saiph, in the lower left-hand region, has an apparent magnitude of 2.1, so that it is slightly fainter than the Pole Star, but its distance from us is thought to be about 2,100 light-years, so that it – like Rigel – is a true 'celestial searchlight'. Gamma, or Bellatrix, and the three stars of the Belt are also highly luminous. And below the Belt may be seen the Hunter's Sword, containing the gaseous nebula Messier 42. It is visible with the naked eye and a superb sight in a moderate telescope. Altogether, Orion lays claim to being the most spectacular constellation in the entire sky.

Last Men on the Moon. It seems quite incredible to think that it was 30 years ago this month that the *last* men walked on the Moon. The final lunar landing of the Apollo programme was Apollo 17, which lifted off from Cape Canaveral on December 7, 1972. The three crew members were Eugene Cernan, Dr Harrison 'Jack' Schmitt and Ron Evans; Schmitt was the first professional geologist to go to the Moon. The landing site was in the Taurus-Littrow region of the Mare Serenitatis, where Cernan and Schmitt touched down in their lunar module on December 11, having left Evans circling the Moon in the command module. Cernan and Schmitt spent a total of 3 days 3 hours on the lunar surface, during which time they completed three EVAs (Extra-Vehicular Activities) of 7.2, 7.6 and 7.3 hours, and covered a total distance of 29 kilometres (18 miles) in their lunar roving vehicle. They also collected 110.5 kilograms of lunar samples. It was Schmitt who found the famous 'orange soil' near the crater that was known unofficially as Shorty. The colour proved to be due to small glassy beads. At the end of the third EVA, Eugene Cernan became the last man on the Moon (so far!) when he re-entered the lunar module, following Dr Schmitt. The crew splashed down safely in the Pacific Ocean on December 19, 1972. The first stage in lunar manned exploration was over.

The Year's End. Today, we are accustomed to regarding December as the 12th month, and December 31 as marking the end of the year, but it has not always been so. Originally the year began in March, and ended with the last day of February. The first Roman calendar contained 304 days, divided into 10 months. This was much too short a 'year', and two more months were added, but the whole system had become confused, and in 44 BC Julius Caesar decided to reform it. The actual work was

Figure 16. Apollo 17 astronaut-geologist Harrison 'Jack' Schmitt standing beside a large lunar boulder. The photograph was taken by mission commander Eugene Cernan, the last man on the Moon.

carried out by the mathematician Sosigenes, but it is always known as the Julian Calendar.

Sosigenes fixed on a year of 365 days, adding an extra day every four years to allow for the fact that the Earth moves round the Sun not in 365 days, but in 365.25. At the same time, the beginning of the year was switched from March to January, so that December became last in the list – but at that time December had only 30 days. April, June, August and October also had 30 days each; February 29, and the rest 31, so that the Leap Year day was justifiably tacked on to February.

The fifth month, formerly known as Quintilis, became 'July' in Caesar's honour. This was all very well, but when Augustus became Emperor of Rome he naturally had to have a month of his own; he took the sixth, Sextilis, which was re-named 'August'. Unfortunately Augustus' month, with 30 days, was now shorter than Caesar's, which had 31. This would not do at all; another day was removed from February and added to August, so that September and November were reduced to 30 days each, with October and December increased to 31. All this political manoeuvring gave the final day of the year as

December 31, which it has remained. The later calendar reform, instigated by Pope Gregory XIII, was merely to drop three Leap Years in every 4 centuries, so making the average length of each calendar year more accurate. Accordingly, a century year is a Leap Year only if it is exactly divisible by 400. Thus 2000 was a Leap Year, but 2100, 2200 and 2300 will not be.

Eclipses in 2002

During 2002 there will be two eclipses, both of the Sun.

1. *An annular eclipse of the Sun on June 10–11* is visible as a partial eclipse from eastern Asia, Indonesia, the Philippine Islands, northern Australia, the Pacific Ocean, most of North America (except the north-east, and southern Mexico). The eclipse begins at $20^h\ 52^m$ and ends at $02^h\ 37^m$. The path of annularity starts in the Celebes Sea, crosses the Pacific Ocean, and then passes just south of Lower California, ending at sunset on the west coast of Mexico, south-west of Guadalajara. Annularity begins at $21^h\ 54^m$ and ends at $01^h\ 35^m$. The maximum duration is 23^s.

2. *A total eclipse of the Sun on December 4* is visible as a partial eclipse from the south-east Atlantic Ocean, Africa, Madagascar, part of Antarctica, the Indian Ocean, Australia (except the extreme east) and Indonesia. The partial phase begins at $04^h\ 51^m$ and ends at $10^h\ 11^m$. The path of totality starts in the south-east Atlantic Ocean and crosses Angola, Zambia, extreme north-east Namibia, northern Botswana, southern Zimbabwe, the extreme north-east of South Africa, and southern Mozambique. It then crosses the southern Indian Ocean before ending in south-eastern Australia. The total phase begins at $05^h\ 50^m$ and ends at $09^h\ 12^m$. The maximum duration is $2^m\ 04^s$.

Occultations in 2002

In the course of its journey round the sky each month, the Moon passes in front of all the stars in its path, and the timing of these occultations is useful in fixing the position and motion of the Moon. The Moon's orbit is tilted at more than 5° to the ecliptic, but it is not fixed in space. It twists steadily westwards at a rate of about 20° a year, a complete revolution taking 18.6 years, during which time all the stars that lie within about 6½° of the ecliptic will be occulted. The occultations of any one star continue month after month until the Moon's path has twisted away from the star, but only a few of these occultations will be visible from any one place in hours of darkness.

There are thirteen occultations of bright planets in 2002, one of Mercury, one of Venus, three of Mars, three of Jupiter, and five of Saturn.

Only four first-magnitude stars are near enough to the ecliptic to be occulted by the Moon: these are Aldebaran, Regulus, Spica and Antares. However, none of these stars undergoes occultation in 2002.

Predictions of these occultations are made on a worldwide basis for all stars down to magnitude 7.5, and sometimes even fainter. The British Astronomical Association has produced a complete lunar occultation prediction package for personal computer users.

Occultations of stars by planets (including minor planets) and satellites have aroused considerable attention.

The exact timing of such events gives valuable information about positions, sizes, orbits, atmospheres and sometimes of the presence of satellites. The discovery of the rings of Uranus in 1977 was the unexpected result of the observations made of a predicted occultation of a faint star by Uranus. The duration of an occultation by a satellite or minor planet is quite small (usually of the order of a minute or less). If observations are made from a number of stations it is possible to deduce the size of the planet.

The observations need to be made either photoelectrically or visually. The high accuracy of the method can readily be appreciated when one realizes that even a stopwatch timing accurate to a tenth of a second is, on average, equivalent to an accuracy of about 1 kilometre (0.6 miles) in the chord measured across the minor planet.

Comets in 2002

The appearance of a bright comet is a rare event which can never be predicted in advance, because this class of object travels round the Sun in enormous orbits with periods which may well be many thousands of years. There are therefore no records of the previous appearances of these bodies, and we are unable to follow their wanderings through space.

Comets of short period, on the other hand, return at regular intervals, and attract a good deal of attention from astronomers. Unfortunately they are all faint objects, and are recovered and followed by photographic methods using large telescopes. Most of these short-period comets travel in orbits of small inclination which reach out to the orbit of Jupiter, and it is this planet that is mainly responsible for the severe perturbations that many of these comets undergo. Unlike the planets, comets may be seen in any part of the sky, but since their distances from the Earth are similar to those of the planets their apparent movements in the sky are also somewhat similar, and some of them may be followed for long periods of time.

The following periodic comets are expected to return to perihelion in 2002, and to be brighter than magnitude +13:

Comet	Year of discovery	Period (years)	Predicted date of perihelion
96P/Macholz (1)	1986	5.2	Jan. 8
2000 WM1 (LINEAR)	2000	?	Jan. 22
7P/Pons-Winnecke	1819	6.4	May 15
22P/Kopff	1906	6.4	Dec. 12

Minor Planets in 2002

Although many thousands of minor planets (asteroids) are known to exist, only a few thousand of them have well-determined orbits and are listed in the catalogues. Most of these orbits lie entirely between the orbits of Mars and Jupiter. All these bodies are quite small, and even the largest, Ceres, is only 913 km (567 miles) in diameter. Thus, they are necessarily faint objects, and although a number of them are within the reach of a small telescope few of them ever attain any considerable brightness. The first four that were discovered are named Ceres, Pallas, Juno and Vesta. Actually the largest four minor planets are Ceres, Pallas, Vesta and Hygeia. Vesta can occasionally be seen with the naked eye, and this is most likely to happen when an opposition occurs near June, since Vesta would then be at perihelion. Below are ephemerides for Ceres, Juno and Vesta in 2002; Pallas is not favourably placed for observation during the year.

1 Ceres

		2000.0 RA		Dec.		Geo-centric distance	Helio-centric distance	Phase angle	Visual magni-tude	Elong-ation
		h	m	°	′	AU	AU	°		°
July	15	1	15.64	− 4	39.7	2.676	2.969	19.9	8.9	96.5W
July	25	1	21.31	− 4	41.6	2.543	2.966	19.4	8.7	104.6W
Aug	04	1	25.20	− 4	55.9	2.415	2.962	18.4	8.6	113.0W
Aug	14	1	27.08	− 5	22.7	2.295	2.959	16.9	8.4	121.9W
Aug	24	1	26.77	− 6	00.9	2.188	2.955	14.9	8.3	131.3W
Sep	03	1	24.17	− 6	48.5	2.097	2.951	12.4	8.1	141.1W
Sep	13	1	19.35	− 7	41.8	2.027	2.947	9.6	7.9	150.9W
Sep	23	1	12.65	− 8	35.6	1.980	2.942	6.8	7.7	159.8W
Oct	03	1	04.64	− 9	23.7	1.960	2.938	5.1	7.6	164.9W
Oct	13	0	56.11	−10	00.2	1.967	2.933	6.1	7.6	162.0E
Oct	23	0	47.98	−10	20.4	2.003	2.927	8.7	7.8	153.6E

1 Ceres (*cont'd*)

		2000.0				Geo- centric distance	Helio- centric distance	Phase angle	Visual magni- tude	Elong- ation
		RA		Dec.						
		h	m	°	′	AU	AU	°		°
Nov	02	0	41.04	−10	21.7	2.063	2.922	11.6	8.0	143.6E
Nov	12	0	35.94	−10	03.9	2.146	2.917	14.3	8.2	133.4E
Nov	22	0	33.02	− 9	28.4	2.247	2.911	16.4	8.3	123.5E
Dec	02	0	32.40	− 8	37.5	2.362	2.905	18.1	8.5	113.9E
Dec	12	0	34.02	− 7	33.8	2.486	2.899	19.2	8.6	104.8E
Dec	22	0	37.71	− 6	19.8	2.617	2.892	19.8	8.8	96.1E
Dec	32	0	43.24	− 4	57.6	2.750	2.886	19.9	8.9	87.9E

3 Juno

		2000.0				Geo- centric distance	Helio- centric distance	Phase angle	Visual magni- tude	Elong- ation
		RA		Dec.						
		h	m	°	′	AU	AU	°		°
Jan	06	9	49.28	+ 0	02.4	1.518	2.317	17.6	8.8	134.6W
Jan	16	9	44.66	+ 0	42.7	1.466	2.344	13.7	8.6	145.5W
Jan	26	9	37.76	+ 1	48.8	1.435	2.371	9.4	8.5	156.7W
Feb	05	9	29.48	+ 3	16.4	1.429	2.399	5.4	8.4	166.7W
Feb	15	9	20.95	+ 4	57.7	1.451	2.427	4.5	8.4	168.8E
Feb	25	9	13.39	+ 6	42.7	1.500	2.455	7.8	8.6	160.4E
Mar	07	9	07.75	+ 8	22.1	1.576	2.484	11.6	8.9	149.7E

4 Vesta

		2000.0				Geo- centric distance	Helio- centric distance	Phase angle	Visual magni- tude	Elong- ation
		RA		Dec.						
		h	m	°	′	AU	AU	°		°
Jan	06	3	47.84	+14	49.0	1.805	2.573	16.4	7.2	132.4E
Jan	16	3	46.12	+15	21.5	1.912	2.572	18.9	7.4	122.0E
Jan	26	3	47.36	+16	01.3	2.031	2.572	20.7	7.6	112.3E

4 Vesta *(cont'd)*

	2000.0 RA		Dec.		Geo-centric distance	Helio-centric distance	Phase angle	Visual magni-tude	Elong-ation
	h	m	°	′	AU	AU	°		°
Feb 05	3	51.32	+16	46.8	2.158	2.571	21.9	7.8	103.3E
Feb 15	3	57.71	+17	36.1	2.289	2.569	22.5	7.9	94.9E
Feb 25	4	06.24	+18	27.4	2.422	2.567	22.6	8.0	86.9E
Mar 07	4	16.61	+19	18.8	2.553	2.565	22.4	8.1	79.5E
Mar 17	4	28.57	+20	08.7	2.681	2.562	21.7	8.2	72.4E
Mar 27	4	41.91	+20	55.8	2.803	2.559	20.8	8.3	65.7E
Apr 06	4	56.41	+21	38.6	2.918	2.556	19.7	8.4	59.3E
Apr 16	5	11.91	+22	16.0	3.025	2.552	18.3	8.4	53.1E
Apr 26	5	28.27	+22	47.2	3.123	2.548	16.8	8.4	47.1E
May 06	5	45.34	+23	11.2	3.211	2.543	15.2	8.4	41.4E
May 16	6	03.00	+23	27.4	3.289	2.538	13.5	8.4	35.8E
May 26	6	21.13	+23	35.3	3.356	2.533	11.6	8.4	30.3E
Sep 23	10	04.54	+14	50.0	3.240	2.446	12.5	8.3	31.9W
Oct 03	10	21.95	+13	30.8	3.157	2.437	14.4	8.3	37.3W
Oct 13	10	39.00	+12	10.5	3.064	2.428	16.2	8.3	42.8W
Oct 23	10	55.65	+10	50.5	2.961	2.419	17.9	8.2	48.5W
Nov 02	11	11.87	+ 9	31.9	2.851	2.410	19.5	8.2	54.2W
Nov 12	11	27.57	+ 8	16.3	2.733	2.400	21.0	8.1	60.2W
Nov 22	11	42.69	+ 7	05.4	2.609	2.390	22.2	8.0	66.3W
Dec 02	11	57.11	+ 6	00.7	2.480	2.381	23.3	7.9	72.7W
Dec 12	12	10.69	+ 5	04.2	2.346	2.371	24.1	7.8	79.4W
Dec 22	12	23.26	+ 4	17.7	2.211	2.361	24.6	7.7	86.3W
Dec 32	12	34.59	+ 3	43.3	2.075	2.351	24.7	7.6	93.6W

A vigorous campaign for observing the occultations of stars by minor planets has produced improved values for the dimensions of some of them, as well as the suggestion that some of these planets may be accompanied by satellites. Many of these observations have been made photoelectrically. However, amateur observers have found renewed interest in the minor planets, since it has been shown that their visual timings of an occultation of a star by a minor planet are accurate enough to lead to reliable determinations of diameter. As a consequence many groups of observers all over the world are now organizing themselves for expeditions should the predicted track of such an occultation cross their country.

Meteors in 2002

Meteors ('shooting stars') may be seen on any clear moonless night, but on certain nights of the year their number increases noticeably. This occurs when the Earth chances to intersect a concentration of meteoric dust moving in an orbit around the Sun. If the dust is well spread out in space, the resulting shower of meteors may last for several days. The word 'shower' must not be misinterpreted – only on very rare occasions have the meteors been so numerous as to resemble snowflakes falling.

If the meteor tracks are marked on a star map and traced backwards, a number of them will be found to intersect in a point (or a small area of the sky) which marks the radiant of the shower. This gives the direction from which the meteors have come.

The following table gives some of the more easily observed showers with their radiants; interference by moonlight is shown by the letter M.

Limiting dates	Shower	Maximum	R.A.		Dec.	
			h	m	°	
Jan 1–4	Quadrantids	Jan. 3	15	28	+50	M
April 20–22	Lyrids	Apr. 22	18	08	+32	M
May 1–8	Aquarids	May 4	22	20	–01	
June 17–26	Ophiuchids	June 19	17	20	–20	
July 15–Aug.15	Delta Aquarids	July 29	22	36	–17	M
July 15–Aug.20	Piscis Australids	July 31	22	40	–30	M
July 15–Aug.25	Capricornids	Aug. 2	20	36	–10	M
July 27–Aug.17	Perseids	Aug. 12	3	04	+58	
Oct. 15–25	Orionids	Oct. 21	6	24	+15	M
Oct. 26–Nov. 16	Taurids	Nov. 3	3	44	+14	
Nov. 15–19	Leonids	Nov. 17	10	08	+22	M
Dec. 9–14	Geminids	Dec. 13	7	28	+32	
Dec. 17–24	Ursids	Dec. 23	14	28	+78	M

Some Events in 2003

ECLIPSES

There will be four eclipses, two of the Sun and two of the Moon.

May 16:	total eclipse of the Moon – Africa, Europe, the Americas
May 31:	annular eclipse of the Sun – North America, Europe, Asia
Nov. 8–9:	total eclipse of the Moon – Africa, Europe, the Americas
Nov. 23–24:	total eclipse of the Sun – Antarctica

THE PLANETS

Mercury may be seen more easily from northern latitudes in the evenings about the time of greatest eastern elongation (April 16), and in the mornings about the time of greatest western elongation (September 27). In the southern hemisphere the corresponding most favourable dates are around February 4 (mornings) and August 14 (evenings).

Venus is visible in the mornings from the beginning of the year until May and in the evenings from November onwards.

Mars is at opposition on August 28.

Jupiter is at opposition on February 2.

Saturn is at opposition on December 31.

Uranus is at opposition on August 24.

Neptune is at opposition on August 4.

Pluto is at opposition on June 9.

Part II

Article Section

Time and Space Near Black Holes

PAUL MURDIN

In fiction, black holes have mystical properties. In the Disney film, *The Black Hole*, a crew of astronauts are drawn into a black hole and appear to go to heaven. In reality, black holes are physical entities just like stars, with distinct scientific properties, and, since we can now see the effects of some of these properties, astronomers are becoming ever more confident that black holes are 'real'.

WHAT IS A BLACK HOLE?

Imagine a body in space, such as a star or planet, and imagine a projectile flung from its surface, like a stone thrown from Earth. If the projectile is thrown at low speed, it rises from the surface and coasts to a halt, falling back to the surface again. The force of gravity from the body has been large enough to counteract the effect of the initial speed of the projectile. There is a speed called the escape velocity, which is just large enough that the projectile just escapes from the gravitational pull of the body – it coasts asymptotically to a halt, only at very large distances. If the projectile is thrown at higher speeds, it definitely leaves the body, and escapes into space.

The escape velocity of a body depends on its mass and size. The escape velocity of the Earth is 11 km/s (7 miles/s), of the Sun 600 km/s (370 miles/s). The smaller a body is in size, the larger its escape velocity because the surface of a small body is close to its centre from which its force of gravity effectively originates. In addition, the more massive a body is, the larger its escape velocity, because a body's force of gravity depends on how much mass it has.

Now imagine a body that has a combination of small size and large mass, so that its escape velocity is very large. It might be such that its escape velocity is larger than the speed of light. In such a case, it would be impossible to throw a projectile into space at the escape velocity, since travelling at over the speed of light is impossible (according to the theory of relativity). A body thrown into space from such a body would always fall back.

This is the basic idea underlying the concept of a black hole. It is a body – a star, or a star cluster, or some such – whose mass and size combine to give an impossibly large escape velocity.

THE FIRST THEORY OF BLACK HOLES

John Michell (1724–93) was a cleric, astronomer and professor of geology at Cambridge. In a lecture to the Royal Society in 1783 he speculated about the effect of gravity on light from the Sun. He based his ideas on the theory that light was corpuscular – small particles, which radiate from the Sun towards Earth. He knew that it was well established from observations of the timing of eclipses of Jupiter's satellites by Rømer in the 17th century that light particles had a speed, measured as 300,000 km/s (186,000 miles/s). He assumed that when the corpuscles were emitted from the Sun, they were emitted at that speed but slowed down as they radiated from its surface. The Sun is of such a size and such a mass that the slow-down effect is small. But Michell calculated if the Sun was 500 times its size and the same density, so that its mass was 100 million times its actual mass, the corpuscles would slow to a halt and not escape from the Sun. In this lecture, Michell grasped the essentials of the theory of black holes, the first appearance of the concept in science. Pierre-Simon Laplace (1749–1827), a director of the Paris Observatory, put forward the same concept in 1795.

THE THEORY OF BLACK HOLES ACCORDING TO GENERAL RELATIVITY

The modern theory of black holes is expressed, not in terms of gravity, the force imagined by Isaac Newton, but in terms of General Relativity as imagined by Albert Einstein. Black hole theory was developed in about 1910 as part of General Relativity by Karl Schwarzschild (1873–1916). In the language of General Relativity, space curves around a massive body due to the gravitational distortion of space time. Light follows curved paths (geodesics) in this space time. If a body is sufficiently massive and sufficiently small, then light from the surface of the body might curve so tightly that it might reach no more than a certain distance from the body. The body would be black, because light would never leave the body – hence the name 'black hole'. In honour of Schwarzschild, this distance is called the Schwarzschild radius of a black hole (see below).

Figure 1. Schwarzschild radius of black holes

	Mass	**Schwarzschild radius**
A galaxy	1 billion Suns	1.5×10^9 km (10 Earth-orbits)
A cluster of stars	1 million Suns	1.5×10^6 km (the solar diameter)
A large star	100 Suns	150 km
The Sun	1 Sun (2×10^{30} kg)	1.5 km
The Earth	6×10^{24} kg	4 mm

Probable black holes include the stars A0620-00 (10 solar masses), V404 Cygni (12 solar masses) and Cygnus X-1 (11 solar masses) and the galaxies M87 (2.4 billion solar masses), NGC 4258 (36 million solar masses) and Sagittarius A* in our own galaxy (2.6 million solar masses). There is obviously quite a range of black holes, so far as their masses go; this hints that there is more than one way to make a black hole.

UNIFIED THEORY OF BLACK HOLES

General Relativity is as good a theory of gravity as we have, and is remarkably successful. Observations of the orbit of the millisecond Hulse-Taylor pulsar show, for example, that discrepancies from General Relativity at most are at the level of 1 per cent. But science works by successive approximation, and there is presumably a better relativity theory than Einstein's. So there is also, probably, a better theory of black holes than Schwarzschild's. It seems that black hole theories come along at intervals of a century, because after Michell and Laplace's theories in about 1800, and Schwarzschild's in about 1900, science is now contemplating Stephen Hawking's theory of black holes which he formulated around 2000. Hawking proposed a theory that unifies General Relativity and quantum mechanics, that is to say gravity and radiation. According to Hawking, radiation can appear spontaneously in the space near a black hole, and, if it appears above the Schwarzschild radius, can radiate away. This decreases the energy of the black hole, and energy moves from inside the Schwarzschild radius to outside. The black hole is no longer black. The creation of Hawking radiation evaporates the black hole.

Hawking radiation is weak. Its wavelength is about the same as the Schwarzschild radius of the black hole, so it is in general very weak – long radio waves. The luminosity of the radiation is about 1 photon every light crossing time of the Schwarzschild radius. The luminosity of a 30 solar mass black hole is about 10^{-31} watts. The total energy of a black hole is $E = mc^2$ so the time for the black hole to evaporate completely can be calculated. For a 30 solar mass black hole, this lifetime is 10^{61} times the lifetime of the Universe. This is therefore not a significant effect for such a black hole, or a bigger one. But if any small black holes were created in the Big Bang, say of 10^{11} kg or smaller (smaller than an asteroid), they would have evaporated by now. Therefore, there are no black holes in space of this size, formed at the Big Bang.

HOW DO YOU MAKE A BLACK HOLE?

Astronomers believe stellar-sized black holes are made by explosions of stars such as supernovae. Chris Kitchin described how black holes might be formed that way in the *Yearbook of Astronomy* 1998, p. 194. Since no stars are more than 100 solar masses, a single star cannot form a black hole like the one in our own Galaxy, with a mass of 2 million suns – a single star does not have enough mass to do that. To form such a massive black hole, a galaxy of stars must be involved. The picture that astronomers are working on at the moment is as follows.

Soon after the Big Bang, galaxies formed from intergalactic clouds of hydrogen, and stars condensed in the newly forming galaxies. The stars moved round the centres of mass of the galaxies, and in orbits that radiated in and out. Some stars orbited in circles, and created galaxies with spiral shapes; stars also accumulated in the centres of the galaxies, gathering into massive dense star clusters. These star clusters of millions or billions of suns formed massive black holes. Nearly all galaxies have central black holes, but some have more massive black holes than others. The Hubble Space Telescope (HST) has pictured numerous galaxies with bright star-clusters (nuclei) at their centres.

The HST has also pictured some active galaxies with extra-bright, extra-condensed nuclei showing jet-like activity and fast-moving material. Some of these galaxies emit X-rays characteristic of hot material. All these phenomena are interpreted as being caused by the black hole feeding on surrounding gas and stars. The material that falls on to the black hole is squashed by the strong gravitational force of the black hole, and is heated (via X-rays). Such a stream of material may fall on to the black hole that the black hole cannot ingest it, and some is thought to get bounced out as a jet. The speed of the fast-moving material in disks around some black holes like M87 has been measured by HST, and interpretation of the orbital speed of the disk reveals the masses of the black holes.

Astronomers have built up a consistent picture, into which these phenomena fit, but there is a worrying feature about the picture – it has little in it that requires the massive object at the centres of galaxies to be a black hole. There is no distinguishing signature here that says that general relativity is involved for example. We can readily understand why this is. The size of the accretion disk that is visible to the HST is at least 0.1 parsec, say of the order of 10 per cent of the distance between the nearest star and us. HST cannot view phenomena in a galaxy of moderate distance that are smaller than this. But the Schwarzschild radius of a billion solar mass black hole is about solar-system sized (see the table above), not inter-star sized. What we can directly see is far from the black hole (say 1000 Schwarzschild radii). We would not expect to be able to view general relativistic phenomena in another galaxy with the HST, or other existing optical telescopes.

If we shift our viewing capacity from light to X-rays, however, we might do better. X-rays are energetic radiation that comes from high temperature material. High temperatures are produced close to the energy source, close to the black hole. There may be characteristics of the X-radiation from the inner disk of material that show effects of general relativity. We would not be able to view the effects directly because X-ray telescopes do not work as well as optical telescopes in showing sharp pictures, but the effects may show on spectral emission from the inner edge of the disks.

SPECTRAL EMISSION FROM A DISK ORBITING A BLACK HOLE

A circular disk of gas orbiting round a star, viewed obliquely, looks like a symmetric ellipse. It produces a spectral emission that contains Doppler-shifted components. The side of the disk that approaches emits blue shifted emissions and the side receding emits red shifted emissions. This produces a spectral emission that is broadened by the disk's rotation. If the rotation is slow, the disk, which is symmetrical, produces a symmetrical emission – the red and the blue components are equal. Of course, if the rotation is slow, the separation of the components is not large.

If the material rotates quickly, however, the symmetry is broken. The high speed of the approaching material produces relativistic beaming. This is a phenomenon of special relativity. Light from the fast-moving material may be emitted isotropically so far as the material is concerned but, viewed by us, the high approach velocity pushes the radiation into a beam pointing towards us. This makes the approaching side of the disk brighter than the receding side of the disk. Therefore, the blue shifted emission is brighter than the red shifted emission.

Other effects of general relativity further break the symmetry. Firstly, the disk no longer appears elliptical, because the back of the disk is viewed above the central black hole. The black hole causes a deflection of the X-rays passing from the rear of the disk, and flips up the back of the image of the black hole.

The black hole also causes a general red shift of emission from the inner edge of the disk. This effect is a time-dilation effect, known as the gravitational red shift. It affects the fast-moving material, since this is the material at the inner edge of the disk. The fast-moving material, whether producing red or blue shifted emission, suffers an overall red shift. This also produces an asymmetry in the spectral emission.

The effect of all this is to change the symmetric, rather narrow spectral emission from a slow rotating disk near a star into an asymmetric, rather broad spectral emission from a fast rotating disk near a black hole. The asymmetry shows the effects of general and special relativity.

Such an asymmetry of spectral emission has been observed in X-rays from several X-ray galaxies. It was first observed in the galaxy MCG-6-30-15 by the Japanese Asca satellite, but shows in active galaxies in general – that is, those that have black hole nuclei. In MCG-6-30-15, the asymmetrical shape of the spectral emission indicates that the disks extend to about 6 Schwarzschild radii, and material is moving at speeds of up to one-third of the speed of light. The shape is a smoking gun of the effects of relativity and directly links general relativity with active galactic nuclei. It is the strongest evidence yet for the reality of black holes.

USEFUL WEBSITES

Ben Bromley's images of an accretion disk around a black hole, with explanations, appear in the following websites:

http://qso.lanl.gov/~bromley/nu_nofun.html
http://cfata2.harvard.edu/bromley/disk_images.html

Making a Deep Impact on UK Government Space Policy

LEMBIT ÖPIK

Probably the most important astronomical thing that happened to me was being born into a family that contained a professional astronomer who believed that the end of the world was nigh. Ernst Julius Öpik was eccentric, strong-minded, passionate, musical – and usually right. His work on the Solar System, mainly at Armagh Observatory and at the University of Maryland in the USA, was often groundbreaking – so much so, that some people regarded his ideas as nuts. Interestingly, I've never met a dinosaur that laughed at my grandad, except perhaps in the Houses of Parliament. The rest of them died out in just the way my grandfather would have predicted. When I was returned as MP for Montgomeryshire, I resolved that I had to maintain an interest in the world beyond party politics, and then I met Jay Tate. Back in 1996 he had begun to meet regularly with the likes of Richard Tremayne-Smith of the British National Space Centre, Dr Jasper Wall, Director of the Royal Greenwich Observatory and Dr David Holland from the Ministry of Defence to talk about the threat posed by asteroids and comets passing near to the Earth. In October 1996, Dr Tom Gehrels and 17 other scientists made contact with 16 politicians and scientists urging UK involvement in Near-Earth Object (NEO) research, and this was followed up with a letter from Dr Edward Teller to the prime ministers of Great Britain and Australia. From these discussions Spaceguard UK was formed. It is an information service for the public, media and the professional – not a political lobbying body.

I first met Jay Tate when he was giving a talk to the Shropshire Astronomical Society in a small community centre in Shrewsbury. By the time he'd finished I knew I had to do something. For the first time in my life, I realized that, with a lot of work and a bit of luck, I could help make the political breakthrough that had so far evaded the Spaceguard programme. I was bound to be regarded as wacky by my colleagues, but, hell, so what? They laughed at Galileo. So this is what I did.

The first step to make the government listen was to get the subject on the agenda. This wasn't easy, since no department is apparently responsible for the end of the world. We got tossed from the Ministry of Defence to the Department of Environment, Transport and the Regions, and even briefly to the Home Office. I know Jack Straw, as Home Secretary, was under pressure about the arrival of illegal aliens, but I though the likelihood of significant influx from the Kuiper Belt was stretching it a bit.

In desperation, we called the House of Commons Library. We explained that the world was in mortal danger of an asteroid impact that could wipe out most of the human race, and most of the rest of life on earth too. We described the heat flash, the blast, tidal waves and electromagnetic pulse that would fry anything that has a microchip in it. Then we asked who should we call. After a few moments' silence, the librarian came back with the immortal reply, 'Well, on the basis of what you've told me, maybe you should be talking to the Archbishop of Canterbury.'

In fact, the high priest of asteroids turned out to be John Battle MP, at the Department of Trade and Industry. So that's where we ended up. That's also when we had our first breakthrough. I managed to secure a 30-minute debate on the floor of the House, under the title Spaceguard. I started with the phrase: 'I've got a problem with asteroids,' which with one fell swoop secured me a lifetime's supply of various creams and lotions from all over the country! Although the debate started around 10.30 at night it generated a pretty big audience from my party and others. I suspect the motive was less about astronomy and more about coming to watch what, many assumed, would be the final speech of my political career.

I told them about the dinosaurs, I told them about Tunguska and I told them about what would happen if a 1km-wide object were inbound to any point in the Atlantic. I described the heat flash and the shock wave, and, funnily enough, they were not laughing when I finished. I think the statistic that upset them most was that you are 750 times more likely to die in an asteroid impact than you are likely to win the National Lottery this weekend.

John Battle MP was the minister who had the great pleasure of responding to this on behalf of the government and he did it well. Against all the odds he chose to take a serious line on it. Perhaps something that helped was that I suspect he was being advised by the same people that I was. Certainly his response was enough to put Spaceguard on the political agenda. He said 'My colleagues and I stand ready to discuss the matter further with him. Our understanding of this subject is developing. The government welcome such positive input. I hope that the honourable gentleman is reassured that the matter can be treated seriously.'

Meanwhile, in another place, in the House of Lords, Lord Tanlaw asked Her Majesty's government: 'What steps are being taken to form a national Spaceguard centre, as part of a European Spaceguard programme, to improve the assessment and probability factor of impact hazard of a near-earth object on the continent of Europe or in the seas surrounding it?' Lord Sainsbury replied positively to the question.

So we were in business at last, and I knew now we had to campaign for a specific outcome that didn't push the government too far in one go. As you may have noticed the government likes task forces because they are cheap and noisy and tell it what it has to do, without it having to do it. But when you are in New Rome, do as the New Romans do. Over a period of three months we pushed the asteroid threat up the agenda. The media, for once, were very helpful. Nigel Nelson of the *Sunday People* was instrumental in keeping up the pressure. Somehow, we managed to package the concept of an asteroid impact in a new way every month for about six months in a row. Usually, this involved finding a close-flying NEO, normally supplied by Mark Bailey at Armagh, and then generating a graphic of what would happen if it hit, say, Birmingham, or anywhere else with a lot of *Sunday People* readers.

Then a miracle happened. In *Armageddon*, Hollywood sent Bruce Willis into space to stop an asteroid from destroying the planet. Together with the film *Deep Impact*, we suddenly had the public's attention. In the sort of irony that makes life so entertaining, Hollywood convinced the UK population that asteroids were a science fact.

In June 1999, we met with Lord Sainsbury, who had taken over as Blair's spokesman on NEOs and we put the case for the Task Force. Nigel Holloway from Aldermaston put across the actuarial risk points, Jay Tate outlined Spaceguard, Mark Bailey frightened everybody with his wild tales of enormous rocks, and Lord Sainsbury made notes. On January 4, 2000, Lord Sainsbury finally announced the NEO Task Force. Eight months later, in the middle of the Liberal-Democrat annual conference, he published the report. That is definitely one of my favourite days in British politics. Even those who had looked down on us from the start were now looking up. And Lord Sainsbury had been very fair. He had delivered a report with 14 clear recommendations that could just help us save the planet.

Then it all went a bit quiet at the government end. So, I started asking questions more loudly again. I raised the issue with the prime minister at Prime Minister's Questions, and in February 2001 I challenged another minister in a space debate to give an update on plans.

I've been suggesting that the next step should be for the prime minister to raise the issue at a forthcoming G8 Summit. After all, the richest nations have the most to lose in terms of infrastructure from an asteriod hit. The cost to these nations would be peanuts. It would cost about a million pounds a year each for ten years to see what's up there, and a total of around £4 billion to divert an asteroid. Put it another way, that works out as 60p a person for a global insurance policy against a pretty messy end.

So what progress am I making in influencing the UK Government space policy?

Firstly a Near-Earth Object Policy can be formulated. Ministers have not committed funds yet, but I think they would be crazy to think they can't. It's so likely that there will be an impact in our lifetimes, of at least a few megatons, that governments should take steps now, rather than respond to public panic the day after. A key recommendation of the Task Force – Recommendation 13 – is that 'A British Centre for Near-Earth Objects should be set up, whose mission would be to promote and co-ordinate work on the subject in Britain; to provide an advisory service to the government, other relevant authorities, the public and the media, and to facilitate British involvement in international activities. In doing so it would call on the Research Councils involved, in particular the Particle Physics and Astronomy Research Council and the Natural Environment Research Council, and on universities, observatories and other bodies concerned in Britain.' If you ask me, to ignore such advice in the ten-pin bowling alley more commonly called the inner solar system would probably turn out to be a serious election loser in the event of a strike.

Secondly, I have learned it's vital to find friends in high places. I had an interest anyway, but the breakthroughs came by influencing decision makers to take the threat seriously, without risking looking foolish.

Thirdly, it takes time. We're still some time away from actually committing money. But I believe it will happen. Minister Alan Johnson, replying for the government said in February 2001: 'Action to increase the detection of Near-Earth Objects with the potential to collide should therefore be addressed internationally. That is what the honourable gentleman has said, and we agree with him.'

And finally, I have discovered that it is still possible to do something that goes beyond the confines of party politics and internal fighting. Sometimes we can reach out for a higher goal, in this case, towards the stars.

I think it would be a pity if, after three billion years or so of evolution, the smartest species in our neck of the suburbs of the Milky Way was extinguished for no better reason than the fact that we were willing to spend hundreds of billions to maintain the capacity to destroy ourselves, but nothing to avert the greatest threat of all. People have different reasons for getting involved in politics. Some do it for fame, others for power over their fellow humans. And some just want to save the world. At this juncture, I hope we will make the decisions to ensure that history will be able to record who was who. And if it works out, I hope my grandad would be proud.

The Big Horn Medicine Wheel

COLIN TAYLOR

Almost 3,000 metres above sea level in the Big Horn Mountains of northern Wyoming, is an irregular circle of stone with spokes radiating from a central cairn (Figure l). Approximately 21 metres (70 feet) in diameter and constructed of flat white local limestone rocks, it was first discovered by Anglo-Americans in the late-nineteenth century. Recognized as a structure of considerable antiquity and mystery, the term 'Medicine Wheel' was coined when describing it.

It is clear that the existence of the Big Horn Medicine Wheel had long been known to those tribes who, in historic times, roamed and hunted in the region.[1] None of these groups, however, claimed to have any detailed knowledge of who built it or why – although they had several legends that attempted to explain its origins.

Now designated as a National Landmark (Figure 2), in recent years both the site and structure have become increasingly important to the Plains tribes and others, and it is revered as a holy place.[2] The tribes claim, and it seems not without considerable justification, that the Big Horn Medicine Wheel encapsulates much of their ancient religious and ceremonial culture, extends back through aeons and, not surprisingly, was frequently keyed to cyclic astronomical events.

The accumulating body of scientific evidence outlined below, which is drawn from the archaeological and anthropological record relating to the North American Indian, is increasingly supporting this view.

THE BIG HORN MEDICINE WHEEL AS A CALENDRIC DEVICE

An association with the Sun was early made by several observers who visited the Medicine Wheel after its discovery by Anglo-Americans in the early 1880s. Local Crow Indians referred to it as the 'Sun's tipi', suggesting perhaps that it was used 'to mark the sunrise'.[3]

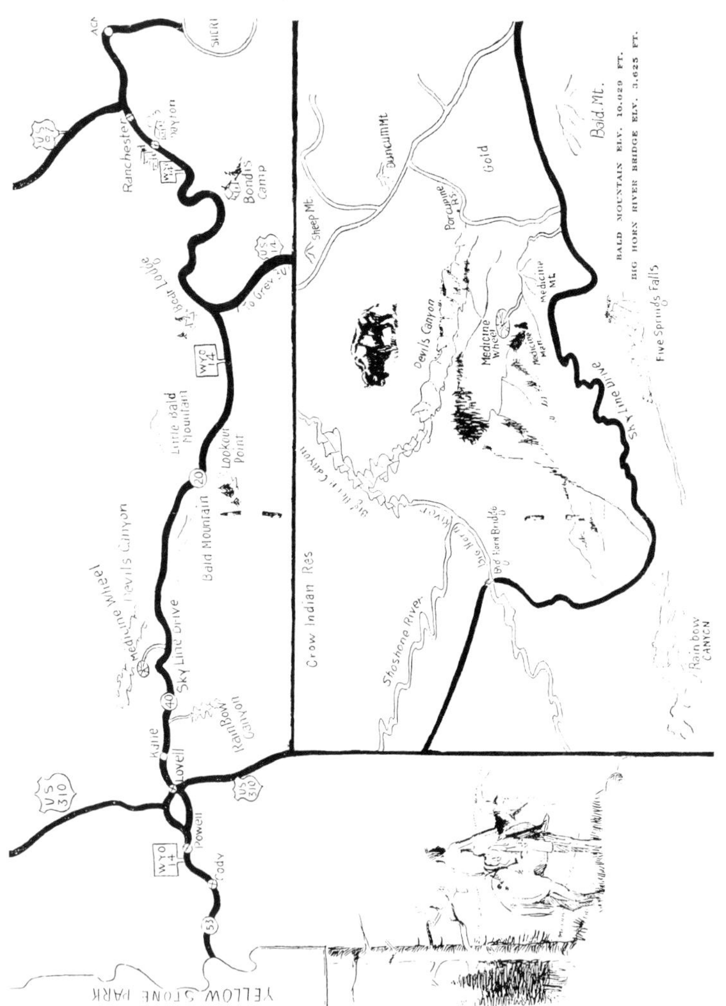

Figure 1. Sketch map showing the Big Horn Medicine Wheel in the Big Horn Mountains of northern Wyoming.

Figure 2. Road sign en route to the Big Horn Medicine Wheel – now designated a National Monument – Wyoming Place No. 22 on the National Register of Historic Places.

It was not, however, until almost 100 years later, in the early summer of 1973, that some careful and detailed astronomical observations and measurements were made on the wheel and cairns.[4] Clearly aware that there was considerable evidence to suggest solstitial alignments of European megalithic monuments – Stonehenge being a prime example – and that Mayan temples and other architectural structures in New Mexico, Colorado, Kansas and Illinois also seemed to have solstitial alignments, this possibility was explored for the Medicine Wheel by the astronomer, John Eddy.[5]

THE SUMMER SOLSTICE

As is well known, the points of sunrise and sunset move daily, such that their paths go through reversals of directions at the times of both the winter and summer solstices. It is then that the Sun both sets and rises at its most southern and most northern positions respectively. Since access to the Medicine Wheel during the winter season was, and still is, virtually impossible, deep snow blocking the long used Indian trail to the plateau, possible use as some sort of observatory for winter solstice was rejected. Even in early summer the trek is often through mountain snow, although the wheel itself is generally swept clean by the wind. This was, perhaps, one of several reasons why the location was chosen since reference points would always be visible during any summer snowfall.

Shortly before dawn at the summer solstice, Eddy positioned himself at the south-western cairn. This was the only one not actually touching the wheel itself; it was, however, connected by an extended spoke some 4 metres beyond the wheel's rim. He then viewed over the central cairn, using it as a foresight. As the Sun rose across the horizon, it was found to be precisely in line with the two points. At sunset, he moved to the south-eastern cairn and again sighted across the central cairn, the Sun was again within close alignment. Eddy observed: 'From backsight E one sees the flash of the dawn sun at the centre of foresight O; from cairn C one sees the last of the setting sun very nearly in line with O.'[6]

After careful consideration of the limitations of certainty of the survey[7], Eddy was led to conclude that the probability, *by chance alone*, of two or six peripheral cairns being aligned to the summer solstice was less than 1 in 4,000. Thus, the case seemed particularly compelling that 'the Big Horn Medicine Wheel [cairns] were built for the specific purpose of marking the summer solstice. Cairns E, O, and C, with but little additional specification of their real reference points, would have permitted their builders to identify the time of the solstice with a precision of several days. Such refinement would have followed quite naturally from repeated annual observation and use.'[8]

HELIACAL RISING OF THE STARS

Eddy then checked the alignments of the remaining four cairns in order to establish a possible link with other astronomical phenomena. In this survey, he was clearly guided by the knowledge that the rising bright stars had often been used by early peoples for calendar references.

In the case of the Big Horn Medicine Wheel, a large number of combinations could be obtained by pairing the six peripheral cairns and so, in order to test for what could be considered reasonably valid astronomical coincidences, it was necessary to employ some restraint. Thus, the use of just one cairn as a common back, or foresight, for more than one other cairn, was used in the quest to identify an alignment with a 'set of significant celestial objects'.[9] It was subsequently found that on using the western cairn (F) as a backsight and then sighting across the northern cairn (A), that Aldebaran, the second brightest (red) star in the sky during the summer solstice, was in alignment. The line from F towards cairn B identified the rising of the blue star, Rigel, while the line towards O defined the rise of the white star, Sirius.

Significant was the fact that these three stars were the brightest in a compact region of the sky that was close to the path of the sun in summer. Furthermore, they rose close to dawn at the only time of the year when access to the site could be reasonably guaranteed.[10] The rising of Aldebaran was of particular interest since, over a period of *at least* 400 years, (approximately 1500 to 1900), it rose on the morning of the 21 June, an hour or more prior to the rising of the sun, thus providing 'the only other celestial signal for the summer solstice'.[11] The same phenomena would then occur with Rigel some 28 days later and with Sirius 56 days later.

THE MOOSE MOUNTAIN MEDICINE WHEEL

These astronomical findings, it should be noted, were quite independent of the circular rim of stones, needing only the centre cairn together with five of the others[12], which leads one to conjecture that the wheel and spokes were later additions and unnecessary for astronomical observations. This was indeed found to be the case with another Medicine Wheel located on the top of Moose Mountain in Southern Saskatchewan. Possibly more ancient than the one in Wyoming, here again five cairns surrounding a large central hub enabled the direction of the sunrise at the summer solstice to be defined as well as marking where the three stars – Aldebaran, Rigel and Sirius – rose.[13] As Eddy observed, 'it bears so strong a resemblance to the pattern of cairns in the Big Horn Wheel that it could have been built from the same set of plans', and he concluded that 'to my mind it confirms, without much doubt, the astronomical use of these mysterious structures . . . [it] meant that for more than 1,000 years the early Indians of the plains were using the same star risings, and the Sun's, to mark the summer solstice.'[14]

John Eddy is foremost an astronomer and although he considered some ethnographic evidence relating to the indigenous people who built and used the Medicine Wheel, he did not pursue this in depth.[15] However, as will be discussed below, the validity of Eddy's tentative conclusions are considerably strengthened by the body of anthropological data, accumulated in recent years, which increasingly demonstrates that those indigenous Americans who were almost certainly responsible for the Medicine Wheel, had a deep interest in, and considerable knowledge of, astronomical phenomena. Clearly then, of no less interest is the identity and motivations of the original architects who laid out the Medicine Wheel structure because much of their ritual, ceremonial, religion and symbolism was traditionally at least 'guided by the stars'.

THE POSSIBLE ARCHITECTS OF THE MEDICINE WHEEL – ETHNOGRAPHIC RECORD

Since some parts of the Wyoming and Saskatchewan Medicine Wheels are obviously very ancient, it is clear that the historic equestrian Plains tribes who roamed and hunted in the Wyoming and Montana area after the acquisition of the horse in the mid- to late-nineteenth century, could not have been the original builders.[16]

Archaeological evidence, however, clearly demonstrates that pedestrian foragers and, later, big game hunters roamed the region for more than 10,000 years, their ancestral homeland being Siberia and beyond.[17] Such transient groups, as has recently been pointed out in the discussion of sacred regions in the Black Hills, tended to adopt such sites 'recognized by their predecessors in the area'. Earlier traditions were subsequently modified to fit both the new physical and conceptual landscapes in which the groups found themselves. 'This processing of borrowing sacred locales and the traditions associated with them took place despite major linguistic, economic, and religious differences among the original and immigrant groups.'[18]

Thus, a most likely scenario for the Big Horn Medicine Wheel is that it was appropriated by a succession of several different groups over an extended period of time. The site of the large central cairn seems particularly ancient, it having yielded lithic materials which date before 2000 BC.[19]

It is little wonder that the site was chosen. At some 3,000 (9,800 feet) metres above sea level, the location is awe-inspiring – just as it must have been for thousands of years. Here, one surveys a largely treeless vista with sweeping valleys and plains below (Figure 3). At night, in the clear atmosphere, there are uninterrupted spectacular views of the sky. Sometimes there is the moan of the wind, at other times complete stillness, while the flat plain is studded with limestone rocks that gleam in the moonlight. All this and its complete isolation clearly endow the area as a place of mystery.

Such sites were sought – probably from time immemorial – for contemplation during the well-known 'vision quest', so important to the North American Indian. In such locations, individuals could savour, more than anywhere else, the perceived spiritual power of the universe. Such religious concepts almost certainly extend back to ancient shamanistic beliefs that can be traced to the earliest homelands of the American Indian – Northern Siberia and beyond. The Cheyenne and others considered that ultimate powers were to be found in the uppermost sky region: the stars, it was said, opened to higher levels of the world above and the constellations were relatives or mythical ancestors who provided signals for actions on earth.[20]

Figure 3. Looking down into the sweeping valleys below the Big Horn Medicine Wheel – a largely treeless vista with the Plains below. The Big Horn River is in the distance. This is the land of the Crow Indians.

A rich knowledge of astronomical lore subsequently evolved from such ancient beliefs. From one group to another, the emphasis changed, but many similar basic teachings were common to many tribes, not least to the Siouan, Caddoan and Algonquian groups who had settled along the border of the eastern Great Plains. Here, for several hundred miles on the banks of the Mississippi and Missouri Rivers, they built large earth lodge villages. Hundreds of such sites have now been located, excavated and mapped.[21]

A number of these villages (dating from *c.* AD 1400 onwards) were fortified and most had a plaza clearly designed to accommodate ceremonial and religious festivals. These ceremonials, particularly some of the most impressive, were subsequently documented in considerable detail by the white observers who came to some of the still extant villages in the early eighteenth century. Later, archaeological studies added to the record.[22] Here, among such tribes as the Pawnee and Arikara (Caddoan), Hidatsa and Mandan (Siouan) and Cheyenne (Algonquian), were complex ceremonials which related to seasonal changes, earth renewal, the calling of the buffalo and other game animals, as well as the harvesting of crops.

These ceremonials clearly had their beginnings which were

embedded in each tribe's own ancient beliefs, although there is some circumstantial evidence that suggests that some were adopted from the ancient Mississippian cultures in the south – perhaps some were even influenced by the Aztecs of Mexico.[23]

In these ceremonials, a quest for power from the sky was frequently sought. Thus, the Pawnees' supreme god, *Tirawahat* – visible through the blue of the sky and the first cause of all – was linked with the world below through *Cu:piritta:ka*, 'female white star' or the 'Evening Star' in the west. Evening Star was conceived as a beautiful woman who had great power over the crops and renewal of the buffalo herds.

The first god placed in the heavens by *Tirawahat*, however, was *Opirikata*, or 'Morning Star' of the eastern sky. *Opirikata* was considered a great warrior who drove all the other stars before him across the sky and it was he whose powers were particularly sought by Pawnee men. It was the union between Morning Star and Evening Star from which the first human being was placed on earth. For this gift of existence, Morning Star was, from time to time, appeased in elaborate ceremonials which included the sacrifice of a young human being – generally a female – in acknowledgement of the tribe's celestial heritage.[24]

These religious concepts, which centred so much around celestial phenomena were extended to the layout of the villages. According to Pawnee belief, each of their villages was founded by a particular star, and owners of sacred star bundles mediated between the sky deities and the community. The earth lodge itself was dedicated in associated building ceremonial not only to the Earth, sky and clouds, but also to the key stars. An elaborate ceremonial for children – the *Hako* – evoked a 'spiritual' sundial. 'As the Sun rises higher, the ray, which is its messenger, alights upon the edge of the central opening in the roof of the lodge . . .' Songs of thankfulness followed as the beam of the Sun moved across the floor. 'The most potent moment of their recitation came as hearth, smoke hole, and Sun aligned in a shaft of light'.[25]

The lodge in which these ceremonials took place was itself considered a sacred place, and when built was dedicated and carefully structured. Thus, an altar within the lodge was representative of the garden of Evening Star. In front of the altar was a deep square hole that represented the resting place of *Tirawahat* in the sky. The lodge faced east so that every morning the Sun and Morning Star poured a beam into the lodge and lit the fire in an act of cosmic procreation. Some of the supporting posts of the earth lodge represented the Evening and

Morning Stars; the North Star – head chief of all the stars – was symbolized in a north post, while the southern post represented the Milky Way, a reference to the flickering campfires of dead Pawnees. It was further considered that the star gods poured down their strength from their appropriate directions in a constant stream of energy through the central opening in the roof of the lodge.[26]

Further to the north among the Mandan villages – in what is now central North Dakota – symbolism associated with their most elaborate ceremonial, the *O-kee-pa*, likewise evoked Sun and star powers. This association is understandable since some Mandans related a legend to account for the original creation of the earth and human population. In this they make reference to certain culture heroes who brought the original 'parents' of the tribe 'down from the sky'.[27]

The *O-kee-pa* ceremonial took place in the summer and commenced at sunrise. Accoutrements carried by one of the main performers, *Numakmax-Ena* or 'Lone Man', made reference to the Sun, Moon and Morning Star and transfer of the so-called turtle drums (key components of the ceremonial) took place at sunset. Two individuals participating on the third day of the ceremonial, were described as representing the night sky. These men had been specially selected by owners of sacred bundles relating to the Moon and stars as many 'mythological characters were believed to be living in the sky as stars'.[28]

As with the Pawnee – and indeed with most of these Missouri River tribes – the Mandan ceremonial lodge was a special design. One distinctive feature was its peculiar flat front and the entrance, a tunnel-like structure, pointed directly towards a cairn in the centre of the village plaza, the whole being likened by one observer to the outline of a gigantic turtle. Of interest is that the Mandan, in common with their neighbours, made a strong connection between buffalo and turtle.[29] It was clearly an important reference to the requirements of every living thing – food and water.

It has long been documented that nearby neighbours of the Mandan – the Hidatsa – made particular reference to the two bright stars, Sirius and probably Rigel.[30] The latter's heliacal rising defined a hand pattern in the sky and signalled the start of their Medicine Lodge ceremony – the *Naxpike*. Offerings made at the time of the *Naxpike* were embellished with representations of Sun, Moon and star symbols, while the pledger of the ceremony carried a hoop to represent the Moon. Hidatsa mythology made references to sky powers and shamanistic journeys of culture heroes across the Milky Way. Such tales were possibly reinforced by a well-documented historical event of a bolide meteor – referred to as a 'sacred arrow' – which disintegrated and exploded over the *Coteau des Prairies*, south of Devil's Lake in present-day North Dakota, towards the latter part of the seventeenth century. It caused massive damage and terror, convincing the surviving ancestors of the Hidatsa that they had been 'inseminated by stars'. The 'hot stone' which actually killed a pregnant Hidatsa woman and hurled her twin foetuses prematurely into the world was truly believed to have come from the far side of the cosmos.[32]

Another important tribe who settled on the Missouri River, but at a relatively late date (*c.* 1730), was the Cheyenne. Of Algonquian linguistic stock for hundreds of years, their historic homeland was far to the east of the Missouri–Mississippi basin, possibly in the region of Hudson's Bay.[33]

The archaeological record and oral history, however, suggest that the ancestors of the Cheyenne were no strangers to the Missouri region and the Plains to the west and north, and that they had travelled through and lived in the region during 'the period from about 500 BC to AD 800'.[34] Further, certain sacred sites (such as present-day Bear Butte in South Dakota) were visited regularly 'at least by ceremonial people [of the tribe]'.[35]

The Eastern and Central Algonquian groups, neighbours of the early Cheyenne, it has been established, had much ceremonial in common with groups in north-eastern Asia.[36] They practised world renewal ceremonials which were set by annual signals from the sky. The Mahicans, Munsees and others of present-day Vermont, for example, used a ceremonial lodge which represented the universe, the constellation Ursa Major being projected upon the floor and three stars in the handle of the Dipper represented three hunters accompanied by a dog (the star Alcor). Furnishings of the lodge, the positioning of the officials within it, the movement and acts of the performers, all corresponded to the position of the stars in the Ursa Major constellation.[37] Important symbolism also centred around the turtle, which was said to carry the marks of the sky space: the turtle was also said to be related to the orientation of the ceremonial lodge.[38] Such ceremonialism was undoubtedly taken with the Cheyenne in their migrations west, when for more than three generations the Cheyenne became firmly associated with the Missouri River earth lodge village complex. Of interest is that they formed a particular friendship with the Arikara.[39] The Arikara were close relatives of the Pawnee who, as already discussed, had ceremonial and mythology that put great emphasis on celestial phenomena.

Thus, rich ceremonialism from the south – with possible influences from Mexico – together with that from the north-east, met at the Missouri River villages, and for more than three generations (in several cases longer), there was interchange of cultural lore and knowledge, not least that relating to astronomical phenomena.

Although these earth lodge villages largely dictated a sedentary lifestyle, there is considerable evidence that some of these people left the Missouri and made sojourns west and across the Plains, not infrequently travelling hundreds of miles. For many thousands of years – as briefly discussed earlier – this region had been occupied by various groups but from about 1400 onwards, the main occupants in the area (now identified as Wyoming and Montana) were the Shoshone.[40] The motivations for excursions from the Missouri were for hunting and trade but some individuals were clearly driven by more intellectual pursuits, such as contemplation and esoteric ritual at sacred sites, which included Bear Butte in present-day South Dakota, and the Medicine Wheel in the Big Horn Mountains of northern Wyoming. This latter site was also known to the Shoshone who told the scholar, Ake Hultkrantz, that it was used for vision quests.[41]

SUN DANCE OF THE PLAINS TRIBES

Well recorded is the importance of sacred sites in the initiation of ritual and ceremonial. Nowhere is this emphasized more than in the Sun Dance of the Plains tribes, a ceremonial in various forms[42] which needed to be performed annually as a very condition of their existence. The Sun Dance incorporated much from the ancient village life including use of astronomical phenomena. The camp circle was symbolic of the constellation Corona Borealis – viewed as a camp circle of the gods above – accoutrements and painted motifs symbolized the Sun, Moon and Morning Star; four participants represented the four world quarters, and they all wore wreaths as symbols of the Sun. There was a hut or tipi of secret preparation which corresponded to a sacred mountain to which the originator of the ceremonial retreated in order to learn the mysteries of the ceremonial. The rites in the secret abode were said to be a reference to the acts performed originally in the sacred mountain – it was a rehearsal to ensure that the ceremonial was correctly performed.[43]

SO WHO BUILT THE MEDICINE WHEEL?

Clearly the human use of the site covers a very extended period of time. A projectile point found in the base of the central cairn of the Big Horn Medicine Wheel was dated to more than 2000 BC and other lithic materials recovered represented a time span of perhaps 5,000 years. Wood from one of the outer cairns was found by dendrochronological techniques to date from about 1760.

In July 1902, S.C. Simms of the Field Museum in Chicago made the first scientific study of the site. Near Lodge Grass, Montana, he sought out and interviewed one of the most famous medicine men of the Crow Indians, 'Sees-the-Living-Bull'. At that time, Sees-the-Living-Bull was about 94 years old; he said that his grandfather spoke of the Wheel as having been built 'by people who had no iron' and that it was a 'dedication to the sun'. A recent consideration of the contents of the interview suggests that Sees-the-Living-Bull 'knew more than he wanted to share with Simms'.[44] The site, it seems, was an important sacred one to the Crow and seemingly its esoteric meaning was not to be divulged easily to relative strangers. Simms' main observations were that the Medicine Wheel bore 'a striking resemblance to the famous Calendar Stone of Old Mexico' and that the smaller huts (or cairns) were, during religious ceremonies, 'occupied by the medicine men'. The larger hut (cairn) in the centre, he reported, was supposed to be the abode of Manitou [the Great Spirit]. The Wheel, he concluded, was of 'great antiquity' being 'carefully and purposely planned': as to the architects, however, he remained mute.[45]

The Cheyenne and the Medicine Wheel

The question of who built the Wheel was taken more seriously in 1922 by the distinguished ethnologist, George Bird Grinnell. He found that among the older Cheyenne, the Medicine Wheel was well known and some of them also referred to the existence of one or more similar sites, as well as pictographs or petroglyphs rendered on cliffs, which were representations of the Medicine Wheel.[46] A much respected, elderly

Cheyenne, 'Elk River', also told Grinnell that the Medicine Wheel was the plan of an 'old-time' Cheyenne Medicine Lodge where the Sun Dance took place. The lines leading towards the centre were representations of the poles of the lodge, and the outer circle of stones, its wall. Elk River also appeared to make reference to the sacred altar, where a buffalo skull would be placed (Figure 4).[47] This virtually faced east towards the entrance and the rising Sun – reminiscent, as discussed earlier, of similar arrangements in the ceremonial lodges of the sedentary Missouri River tribes and consistent with Eddy's astronomical observations of the summer solstice.[48] Grinnell's detailed report, which also made reference to interviews with Sioux Indians, mythology and his own personal experiences with the Cheyenne, led him to conclude that the Big Horn Medicine Wheel had been built by the Cheyenne.[49]

That the Cheyenne were the architects not only of the Big Horn Wheel but also of the Moose Mountain Wheel in Saskatchewan has been taken up in recent years by the specialist on the Cheyenne, Karl Schlesier. Pivotal to Schlesier's hypothesis is that the ancestors of the Cheyenne were early occupants (500 BC to AD 800) of the Northern Plains and that they brought to the region shamanistic beliefs which can be traced back to northern Siberia. These proto-Cheyenne, he suggests, built the Wheels to be directed to the 'spirits of the world above'. The positions of the cairns were oriented to define the direction of the heliacal risings of the three stars Aldebaran, Rigel and Sirius of summer dawn, such stars being of vital importance in proto-Cheyenne religious philosophy. The proto-Cheyenne subsequently abandoned the Plains region moving east and leaving behind the stone monuments. Some individuals, however, travelled back from time to time to these sacred sites, the link with them never being fully severed.

Nearly a thousand years later, in the eighteenth century, the Cheyenne came back to the Plains region. Finally, as equestrian nomads in the 1780s, they roamed the Plains and according to Schlesier's hypothesis, made use of the Big Horn Medicine Wheel for one of their most important ceremonials – the *Massaum*. Here, Schlesier identifies a ceremonial period which consisted of 56 days. It commenced with the heliacal rising of the red star, Aldebaran 'around June 22, on the same day or one day before summer solstice. This star flashed briefly in early dawn above the north-eastern horizon. It ended 56 days later when the white star – Sirius (Alpha Canis Majoris) – rose heliacally in the south-east. The *Massaum* was annually held in the middle of this period. The

Figure 4. The sacred altar mentioned by the elderly Cheyenne, 'Elk River', with a buffalo skull in the foreground. At the summer solstice, sunlight would strike this.

fifth, public day of the ceremony was tied to the heliacal rising of the blue star of summer dawn – Rigel (Beta Orionis) – that flashed across the horizon from the south-east. Rigel appeared for the first time exactly 28 days after Aldebaran's first rising and was in turn followed by the rising of Sirius 28 days later.'[50]

By reference to Cheyenne informants, Schlesier was able to key in the movements of the players in the ceremonial which then focused on the blue flash of Rigel. This started the sacred hunt. Blue paint was used to embellish the sacred buffalo skull and with Maltese Cross design – this made reference to Rigel. A Red Wolf figure represented Aldebaran and a White-Horned Wolf figure, Sirius.

While Schlesier's references to the use of these stars in the *Massaum* is convincing, the use of complex star symbolism, as has been explained, was certainly not exclusive to the Cheyenne in the Plains Indian ceremonial round. From the Pawnee in the south, to the Mandan in the north on the Missouri River, commencement of ceremonial and the sequence of subsequent ritual was paced by the sunrise, sunset and movement of the stars.

In the historic period, the Big Horn Medicine Wheel was firmly in the domain of the Crow who now, as equestrian nomads, were constantly at war with the Cheyenne. It seems inconceivable that the latter tribe would camp in the valleys adjacent to the Big Horn Wheel in order that the ceremonial could be paced by the Cheyenne holy men from the sacred mountain. Interesting and compelling though Schlesier's hypothesis is, his explanations have caused considerable controversy. This is mainly because several of his claims are based on an interpretation of esoteric shamanistic language, the meaning of which has been revealed only to him, by Cheyenne specialists. Furthermore, the archaeological evidence he puts forward tends towards personal assertion rather than demonstration by accepted facts. The case for the Cheyenne being the original architects of the Medicine Wheels is still, therefore, to my mind, open to question.[51] However, considering their cultural history, the astronomical knowledge of the early Cheyenne and its important influence on ceremonial and ritual is not disputed.

The Crow Indians and the Big Horn Medicine Wheel

For a more convincing explanation that is certainly more strongly supported by considerable archaeological and historical evidence, we should look in the direction of the Crow Indians – more specifically, at one group of that tribe, the Mountain Crow. There is little question that the Crow traditionally used the site of the Medicine Wheel for vision quests. For example, the anthropologist, Robert H. Lowie, reported in the 1920s that 'Many of the Crow would go there to fast; the structure has been there as long back as any period alluded to by previous generations.'[52]

As was related earlier, Simms' interviews with Sees-the-Living-Bull and other Crow informants show that either they were reluctant to share esoteric knowledge or that they simply did not know – the wrong people were interviewed.[53]

Most explanations relating to the Medicine Wheel have tended to be strongly influenced by the much respected anthropologist, Dr George Bird Grinnell who, as recorded earlier in this article, visited the Big Horn site in the 1920s. Its layout was then likened to a Cheyenne Medicine Lodge. Unfortunately, such an observation has only limited validity because most Medicine Lodges of the Plains Indians were actually all laid out in a similar fashion – although there was one notable exception which is relevant to this debate and which will be discussed shortly.

The Crow originally lived with the Hidatsa in the numerous earth lodge villages along the Missouri River. In about 1600, however – possibly earlier – one group abandoned horticulture and became exclusively pedestrian nomads (Figure 5). To the west were great herds of grazing animals, buffalo and antelope, the changing climatic conditions at that time ensuring increased moisture and lush pasturage. Although moving into a domain traditionally occupied by several Shoshonean groups, the region was so vast and rich that the increase in demand on resources caused little conflict. Indeed, the Mountain Crow – as the group became known to history – formed close friendships with the Shoshone and in later years there was considerable intermarriage.[54] With these Mountain Crow were several families of Mandan, a tribe particularly well known for their rich ceremonialism, whose cultural history extended to the south where astronomical phenomena were of great significance to the culture.[55] This combined group left at a time when the complex calendrical ceremonials and ritual were well developed, the importance of which can hardly be overestimated. Thus, when discussing the Hidatsa and Mandan, George Catlin recorded of the important *O-kee-pa* ceremonial that they attributed to it 'not only their enjoyment in life, but their very existence; for traditions ...

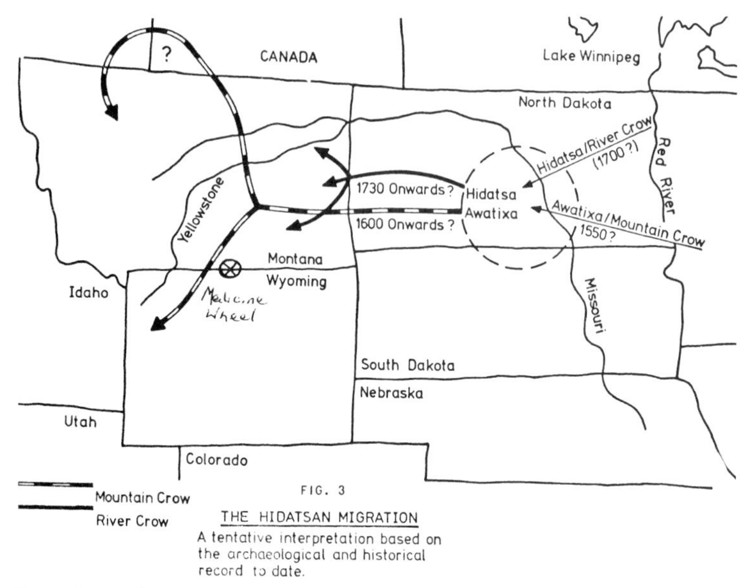

Figure 5. Possible migration paths of the Mountain Crow (and some Mandan) from the Missouri River to present day to Montana and Wyoming, c. 1550–1600. Note that the route passes close to the Medicine Wheel site.

instructed them in the belief that the singular forms of this ceremony produced the buffaloes for their supply of food, and that the omission of this annual ceremony . . . would bring upon them a repetition of the calamity which their traditions say once befell them, destroying the whole human race, excepting one man.'[56]

The Mountain Crow and their Mandan companions thus left behind a rich religious inheritance and the move to pedestrian nomadism clearly limited their ability to carry the complex religious paraphernalia – the so-called Medicine Bundles – with them in order to continue the ceremonials.

This makes the link between the Big Horn Medicine Wheel and the Mountain Crow of particular significance. There was a need for some calendrical structure so that the annual ceremonials, as practised on the Missouri River, could be continued. As one scholar has observed, 'some ceremonies were left behind'.[57] Probably, in particular, were those which were to do with horticulture but others that were directed towards culling and improving the game, were retained. In such ceremonies there was 'interchange between various [mythological] characters and the Sun, the Moon, and several stars, who are intermarried or linked in some way.'[58]

Friendship with the Shoshone would have acted as an incentive to travel to the sacred Big Horn Mountain site which, from ancient times, the archaeological record suggests (particularly the central cairn) had been employed for the vision quest by the indigenous inhabitants. The site was then modified, and other cairns or huts built to accommodate the need for summer solstice observations and viewing of the subsequent heliacal rising of the stars – traditions which can, as discussed before, be related to earlier days on the Missouri.

Of great interest is that the Medicine Wheel's circle of stones is actually flat-fronted and from that flat front is an extension of stones to an outlying cairn. The orientation of this extended axis is to the south-west (Figure 6). Early Mandan sites in North Dakota invariably have a ceremonial lodge oriented towards the south-west and central supporting posts 'appear to have been oriented with the summer soistice'.[59] Traditionally, as mentioned before, as a reference to earlier sacred structures, the Mandan ceremonial lodge was built in a semicircular form but with a flattened front.

The Big Horn Medicine Wheel also has a turtle-like configuration – four of the cairns being the legs, one for the tail and the extension for the head (Figures 6 and 7). Such a shape, it has been suggested, was used by the Mandan in the building of their sacred lodge which was used for one of their most important ceremonials, the *O-kee-pa*, part of which made reference to the calling of game to the village.[60]

The Big Horn Wheel location would probably not have been conducive to tribal gatherings. It is some distance from water and the sloping rocky terrain is unsuitable for tipi dwellers. However, the magnificent surrounding valleys and river courses would have afforded the necessary encampment site for a summer gathering. From there, religious leaders could make recourse to the Big Horn Medicine Wheel. Here, the necessary preparations and observations could be carried out at and near the summer solstice, signals from the sky pacing, as Schlesier suggests was the case for the Cheyenne, the ceremonials that took place in the valley below.

With the passage of time, the importance of this link with Missouri village ceremonialism progressively weakened and the site, although still revered, was abandoned for large gatherings. By the mid-nineteenth

Figure 6. View of the Big Horn Medicine Wheel showing 28 spokes and cairns. The aerial view suggests – and this is reinforced by Mandan mythology – a turtle configuration.

century, such impressive ceremonials as the *O-kee-pa* and the *Massaum* had all but ceased. With them went the detailed knowledge of the heavens kept by religious specialists who employed it to ensure, by correct ceremonial, survival of the tribe. The new generation of equestrian nomads, although still recognizing echoes from the past, now viewed the vault of the sky in a somewhat different way.[61]

At the Medicine Wheel, however, is encapsulated without doubt so much of the rich ceremonialism of the early indigenous Americans, a good part of which, from time immemorial, was 'guided by the stars'.

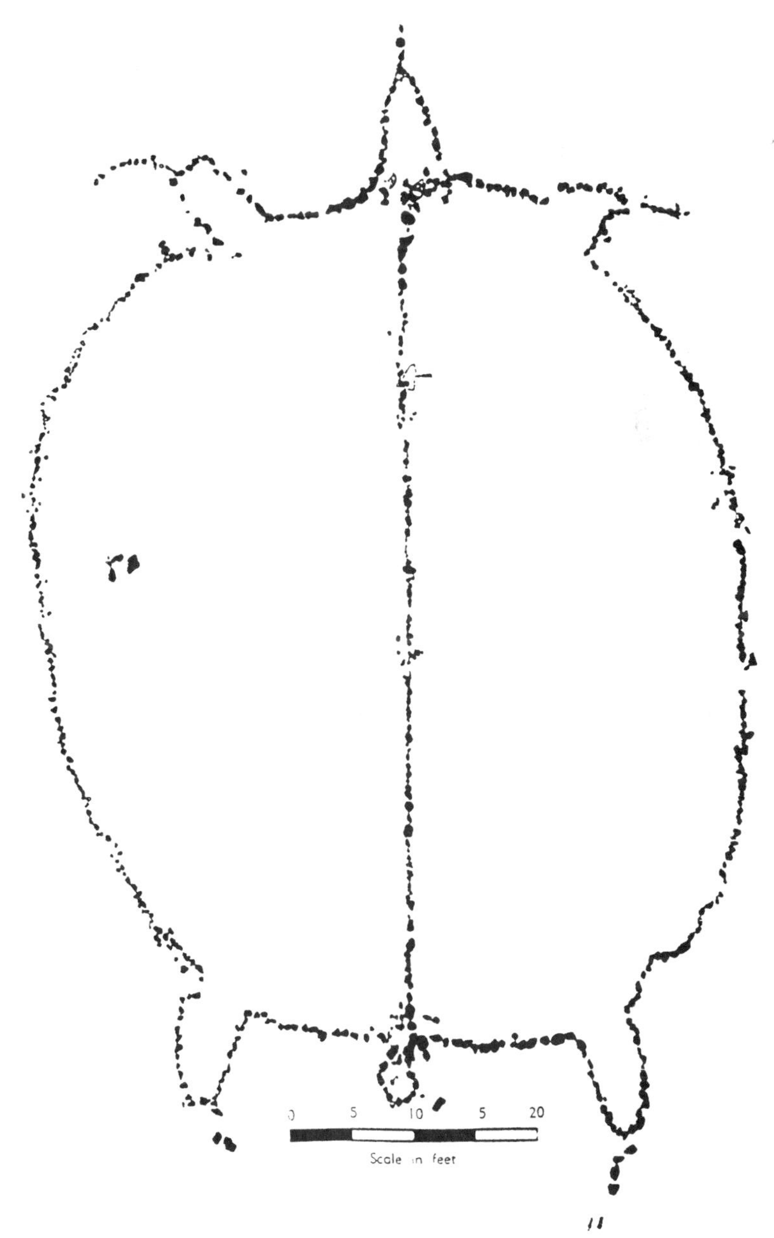

Figure 7. Turtle effigy on the Saskatchewan Plains.

NOTES

1 The Shoshone, Crow, Cheyenne and Arapaho.
2 The site is under the auspices of the National Parks Service and a wire fence now protects the Medicine Wheel itself. Together with a number of interested scholars and Plains Indian representatives, I was recently allowed into the enclosure (September 6, 2000). We were led by John Hill, a senior Crow Indian, who has long been associated with the site.
3 Eddy, 1974, p. 1036.
4 These were carried out by John Eddy, then on the staff of the High Altitude Observatory, National Centre for Atmospheric Research, Boulder, Colorado. Recently (January 2001), I spoke to Dr Eddy about his research. After more than 30 years he was still convinced that the Medicine Wheels, both the Big Horn and Moose Mountain, had the astronomical associations that he had so carefully documented in the 1970s.
5 Several contemporary Pueblo and Apache ceremonials are also timed by the stars. See Farrer, 1989, pp. 223–36.
6 Eddy, 1974, p. 1037.
7 Eddy speculated (with some justification based on previous archaeological evidence) on the possibility of the central cairn originally having a post set in it that would have acted as a more accurate gnomon.
8 Eddy, 1974, p. 1038.
9 *Ibid.*, p. 1039.
10 Other combinations were also considered which included Betelgeuse and Capella but for several reasons these were rejected. Capella, for example, could not be marked on the horizon at the latitude of the Big Horn Mountains.
11 Eddy, 1974, p. 1039.
12 Cairn D could not be satisfactorily accounted for, on astronomical grounds, but can possibly be explained by reference to the historical record, as a completion of a turtle image.
13 This was after discussions with the late Dick Forbis, Professor of Archaeology at the University of Calgary, as well as with Alice and Tom Kehoe, formerly the Provincial Archaeologists in Saskatchewan. John Eddy also conferred with Dr Hugh Dempsey of the Glenbow Foundation in Calgary, who was clearly impressed by Eddy's careful,

detailed scientific and objective stance in his study of Medicine Wheels (communication from Hugh Dempsey to the author, January 2001).

14 Eddy, 1977, p. 146. Dr Eddy's travels across the Canadian Plains led him to observe that many of the rock cairns had simply been built as 'statements' that a group had passed through the area. Only Moose Mountain was found to have astronomical associations (discussion with Eddy, January 2001). Hugh Dempsey's studies of the Blackfeet suggest that some of these stone monuments were to dead high-ranking Blackfeet warriors. See Dempsey, 1956, pp. 177–82.

15 For example, the early peopling of the region by the Mountain Crow accompanied by some Mandan families (see 'The Crow Indians and the Big Horn Medicine Wheel', the last section in the text).

16 The acquisition of the horse and its impact on Plains culture is discussed in Taylor (1994).

17 Petroglyphs, pictographic images and ancient campsites document the evidence for this. See Wedel, 1961 p. 280; Keyser, 1987; Schlesier (ed.), 1989.

18 Sundstrom, 1996: 187. Many of these sites, it should be noted, have astronomical associations which were documented, for example, by the Lakota on star and landscape maps – although few fully understood their meaning.

19 Grey, 1962, p. 317.

20 Schlesier, who has made a detailed study of the Cheyenne, has convincingly demonstrated links between the northern Siberian cultures and this important Plains Indian tribe. Similar concepts were almost certainly held by other Algonquian groups. Schlesier, 1987, p. 46.

21 Wedel, 1961; Howard, 1962; Wood, 1967.

22 Catlin, Ewers ed., 1967; Wood, 1967; Taylor, 1996.

23 See Linton (1922) and Chamberlain (1982).

24 The Pawnee leaders found this a most distasteful part of the ceremonial. See Taylor, 1994, pp. 67–74.

25 Nabokov and Easton (1989).

26 *Ibid.*, pp. 138–9.

27 Bowers, 1965, p. 21.

28 Bowers, 1950, p. 131.

29 Certain meteorites were associated with a turtle, and tribes such as the Cheyenne and Lakota documented on painted hides images of turtles in the sky, formed by various star formations (for the Lakota, see Goodman [1992]).

30 Matthews 1877, p. 153.
31 Bowers 1965, pp. 313 and 319.
32 This is based on a communication with the scholar, Mike Cowdrey (January 2001).
33 Grinnell, 1923, Vol. I, p. 4.
34 Schlesier, 1987, p. xiii.
35 *Ibid.*, p. 138.
36 *Ibid.*, pp. 45–9.
37 Speck (1945: 32).
38 See Schlesier, 1987: particularly, pp. 175–9.
39 Grinnell, 1923, Vol. I, p. 9.
40 Hultkrantz (1968).
41 Discussion with the Shoshone specialist, Professor Ake Hultkrantz (January 2001).
42 Schlesier (1990).
43 This concise analysis of the Sun Dance is based on Dorsey, Hodge ed., 1910, p. 651.
44 Richard Edwards (Professor of Law, University of Toledo and a scholar of the Crow) communication with the author. (January 2001). Professor Edwards kindly made Simms' original field notes available. See also Simms (1903).
45 *Ibid.*, 1903, p. 107.
46 This is an interesting observation. As was discussed earlier, the astronomer, John Eddy, found a Medicine Wheel at Moose Mountain, which has close similarities to that in the Big Horn Mountains.
47 Grinnell, 1922, p. 302.
48 Of interest is that when Simms had visited the Medicine Wheel some 20 years earlier, as discussed in the text, he had found a buffalo skull in this position – which he promptly removed as a souvenir! This specimen is now in the Field Museum, Chicago.
49 The interested reader is referred to Grinnell, 1922, pp. 299–310 for his full report on this site.
50 Schlesier, 1987, p. 84.
51 A critical review of Schlesier's work is in *Plains Anthropologist*, 1990, pp. 35–127.
52 Lowie, 1922, p. 436.
53 Loendorf, 1986, p. 26.
54 (a) For a detailed analysis of this Crow migration, see Taylor (1984).
(b) Of the various groups comprising the Hidatsa, it was the Awatixa

from whom the Mountain Crow split and these people were in turn closely associated with the Mandan.

55 Bowers, 1965, p. 21.

56 Catlin, Ewers ed., 1967, p. 39.

57 Loendorf, 1986, p. 27.

58 *Ibid.*, 28.

59 *Ibid.*

60 (a) I have earlier discussed the structure and symbolism of the Mandan ceremonial lodge in detail (Taylor [1996]).
(b) Note that this combined quadruped and turtle association is a symbolic reference to two vital resources for human life – food and water.

61 (a) This was the subject of my earlier paper on Plains Indian Astronomy (Taylor, 2000, pp. 147–61).
(b) See also McCleary (1997) for a recent review of Crow Indian astronomy and knowledge.

BIBLIOGRAPHY

Bowers, Alfred W. (1950), *Mandan Social and Ceremonial Organization*. The University of Chicago Press.

Bowers, Alfred W. (1965), 'Hidatsa Social and Ceremonial Organization'. In *Bureau of American Ethnology Bulletin* **194**. Smithsonian Institution.

Catlin, George (1967), *O-kee-pa: A Religious Ceremony and other Customs of the Mandans*. Edited by John C. Ewers. Yale University Press.

Chamberlain, Von Del (1982), *When Stars Came Down to Earth: Cosmology of the Skidi Pawnee Indians of North America*. University of Maryland, Center for Archaeoastronomy.

Dempsey, Hugh (1956), 'Stone Medicine Wheels – memorials to Blackfoot war chiefs'. In *Journal of the Washington Academy of Sciences*. **46**, No. 6: 177–82.

Dorsey, George A. (1910), 'Sun Dance'. In *Handbook of American Indians North of Mexico*. 2 Vols. Edited by Frederick Webb Hodge. Bureau of American Ethnology Bulletin 30. Smithsonian Institution.

Eddy, John A. (1974), 'Astronomical Alignment of the Big Horn Medicine Wheel'. In *Science*. **184**: 1035–43.

Eddy, John A. (1977), 'Probing the Mystery of the Medicine Wheels'. In *National Geographic.* **151**: 140–6.

Farrer, Claire (1989), 'Star Clocks: Mescalero Apache Ceremonial Timing'. In *Amerindian Cosmology, Cosmos 4, Yearbook of the Traditional Cosmology Society.* Edited by Don McCaskill. *The Canadian Journal of Native Studies.*

Goodman, Ronald (1992), *Lakota Star Knowledge: Studies in Lakota Stellar Theology.* Sinte Gleska University.

Grey, Don (1962), 'Summary Report of the Medicine Wheel Investigation'. 17th Annual Field Conference. *Wyoming Geological Association Guidebook*: 316–17.

Grinnell, George Bird (1922), 'The Medicine Wheel'. In *American Anthropologist.* **24**, No. 3.

Grinnell, George Bird (1923), *The Cheyenne Indians: Their History and Ways of Life.* 2 Vols. Yale University Press.

Howard, James (1962), 'Report of the Investigation of the Huff Site 32 MO11 Morton County, North Dakota 1959'. Anthropology Paper No. 2. University of North Dakota.

Hultkrantz, Ake (1968), *Shoshoni Indians on the Plains: An Appraisal of the Documentary Evidence.* Zeitschrift für Ethnologie, Band 93, Heft 1 u 2.

Keyser, James D. (1987), 'A Lexicon for Historic Plains Indian Rock Art: Increasing Interpretive Potential'. In *Plains Anthropologist*, **32**, No. 115.

Linton, Ralph (1922), *The Thunder Ceremony of the Pawnee.* Field Museum of Natural History.

Loendorf, Larry L. (1986), 'Remnants of the Mountain Crow'. In *Mountain People.* Edited by Michael Tobias. University of Oklahoma Press.

Lowie, Robert H. (1922), 'Religion of the Crow Indians'. Anthropological Papers. 25. Part 2. American Museum of Natural History.

McCleary, Timothy P. (1997), *The Stars We Know: Crow Indian Astronomy and Lifeways.* Waveland Press, Inc.

Matthews, Washington (1877), 'Ethnography and philology of the Hidatsa Indians'. Miscellaneous Publication No. 7. US Geological and Geographical Survey.

Nabokov, Peter and Easton, Robert (1989), *Native American Architecture.* Oxford University Press.

Schlesier, Karl H. (1987), *The Wolves of Heaven.* University of Oklahoma Press.

Schlesier, Karl H. (1989), *Plains Indians, A.D. 500–1500.* University of Oklahoma Press.

Schlesier, Karl H. (1990), 'Rethinking The Midewiwin And The Plains Ceremonial Called The Sun Dance'. In *Plains Anthropologist* 35–127.

Simms, S. C. (1903), 'Traditions of the Crows'. *American Anthropology.* Vol. 11. Field Columbian Museum.

Speck, Frank G. (1945), 'The Celestial Bear Comes Down to Earth'. Scientific publications. No. 7. Reading Public Museum and Art Gallery.

Sundstrom, Linea (1996), 'Mirror of heaven: cross-cultural transference of the sacred geography of the Black Hills'. In *World Archaeology.* **28** (2): 177–89.

Taylor, Colin (1984) 'Crow Rendezvous'. In *Crow Indian Art,* pp. 33–48. Edited by D. and R. Lessard. Chandler Institute.

Taylor, Colin (1994), *The Plains Indians.* Salamander Books Ltd.

Taylor, Colin (1996), *Catlin's O-kee-pa: Mandan Culture and Ceremonialism. The George Catlin O-kee-pa Manuscript in the British Museum.* Wyk auf Foehr: Verlag für Amerikanistik.

Taylor, Colin (2000), 'Astronomy of the Plains Indians'. In *2001 Yearbook of Astronomy,* pp. 147–61. Edited by Patrick Moore. Macmillan.

Wedel, Waldo R. (1961) *Prehistoric Man on the Great Plains.* University of Oklahoma Press.

Wood, W. Raymond (1967), 'An Interpretation of Mandan Culture History'. *Bureau of American Ethnology Bulletin 198.* Smithsonian Institution.

ACKNOWLEDGEMENTS

I have combined my knowledge of North American Indian ethnology with that of astronomy to give what I hope is a measured assessment of the Big Horn Medicine Wheel and how its astronomical content – as so ably documented by John Eddy more than 30 years ago – fits in with the Plains Indian lifeway.

In order to update and recheck much of the published work on the Medicine Wheel, I contacted many specialists in astronomy, archaeology, ethnology and cultural history. I am particularly grateful to Stu Conner of Billings, Montana, who made his entire file on the Medicine Wheel available to me. Others who were generous with their time were: Kolleen Bean, John Brumley, Winfield Coleman, Mike Cowdrey, Joe Medicine Crow, Hugh Dempsey, John Eddy, Ken Feyhl, John Hall, Ake Hultkrantz, Dietmar Kuegler, Larry Loendorf, Imre Nagy, Karl Schlesier and Dr K. A. Wipf.

Extrasolar Planets and the Upsilon Andromedae System

DAVID WELDRAKE

There have been few discoveries in the history of astronomy that have caught the imagination as much as the discovery of the first extrasolar planet. That planet was discovered orbiting the single main sequence star 51 Pegasi in 1996 (Mayor & Queloz, 1995, Marcy *et al.* 1997). The location of this world (only 0.05AU from the parent star) and its likely makeup and conditions are different from any planet previously studied, and the discovery has produced to a new field of astrophysical research. Since antiquity, extrasolar planets have been predicted to exist, the first evidence being the observation of a dust-shell surrounding the star Beta Pictoris in the 1980s. The 51 Pegasi discovery proved to be the first of many.

The first actual 'system' of planets (multiple bodies surrounding a single parent body) was found orbiting a 6.2 millisecond pulsar PSR 1257+12 (Marcy & Butler, 1998, Wolszczan & Frail, 1992, Wolszczan, 1992). This led to a great deal of speculation due to the destructive power of the supernova explosion that produced the pulsar in the first place. The existence of planets in a system with a history as violent as this suggests that planetary formation is a very robust process, and this discovery provides a clue to the diversity of plausible planetary systems. This system is currently the only one (at the time of writing) to possess multiple terrestrial-mass bodies. Perhaps they are the cores of once massive planets, all but blown away by the supernova, or did they form after that event?

METHOD OF DETECTION

The detection of extrasolar planets involves a lengthy study of the proper motion of the parent star. The planets themselves are far too small and faint to be directly imaged, at least with current technology, and their presence is inferred from astrometric studies of the star itself. All stars have a measurable motion through space, the so-called 'proper motion', and this traces out a straight line across the sky when followed for many years. If, however, the star is orbited by an unseen massive body, the gravitational attraction of that body will 'pull' the star around their common centre of mass. The result is observed as a wobble in the proper motion, and a study of this wobble allows the mass and orbital period of the unseen body to be determined fairly accurately.

Planets of higher mass will subsequently have a greater gravitational effect on the star, and hence the majority of planets currently known have masses greater or equal to that of Jupiter. However, as detection technologies improve, bodies of mass less than this, and possibly with masses as small as that of the Earth, will undoubtedly be found. Indeed quite recently planets of Saturn-mass have been detected (about 0.3 Jupiter mass), and it is expected that the first terrestrial-mass body will be discovered around a solar-type star within the next few years.

Other methods of detection have also been used in recent years. A particularly successful one concerns the study of the brightness of the parent star (photometry). If an orbiting planet passes across the face of the star as seen from Earth, a transit will occur, and if of sufficient size, it follows that the apparent brightness of the star will drop. The amount of brightness drop would be very small, around 0.01 magnitude, but nevertheless should be detectable with current technology. Indeed, in 1999, the seventh magnitude star HD209458 in Pegasus was observed to sport such a drop in brightness (after being predicted to be orbited by a 51 Pegasi-type planet). The resultant analysis allowed the planet's density value to be estimated. It turned out to be only 380 kg/m^3. This makes the planet much more rarefied that either Jupiter or Saturn. Such a low density is to be expected from these hot Jupiter-type planets (of which HD209458 'B' is an example), which experience extreme temperatures (about 1400 K), and would be significantly bloated up by the heat.

OTHER MODELS TO ACCOUNT FOR THE OBSERVATIONS

There are a couple of alternative hypotheses which can account for the observed Doppler periodicities without needing the presence of a 'hot

Jupiter', 51 Pegasi-type planet. These concern the features associated with the star itself, and these ideas have to be considered when a stellar periodicity or transit is suspected.

1. *Stellar Pulsations*
The pulsations of a star (i.e. Cepheid variables, Mira-type variables, etc.) could account for the measured wobble in the star's radial velocity. motion. However, pulsations on a scale necessary to produce a periodicity of the magnitudes observed would cause the brightness of the star to change by quite an appreciable, and hence easily detectable, amount. None of the stars thought to harbour 51 Pegasi-type planets display such magnitude variations.

2. *Starspots*
The presence of a substantial starspot on the surface of the star can account for the drop in brightness which would otherwise be attributed to a planetary transit. Such large starspots have been observed on red supergiant stars such as Betelgeux, and can occur on solar-type stars if the magnetic field is sufficiently large; this is a particularly important factor when dealing with contact binary systems (ER Vulpeculae being a good example). However, to account for the HD209458 transit, a starspot must move across the disk at speeds equal to the calculated orbital period of the planet (four days). Such a rapid rotation of the star would cause a significant broadening of the stellar absorption lines, which are not observed in the case of HD209458.

These ideas can be disregarded in the discovered systems, and the evidence is highly supportive of there being a planet orbiting these stars. This raises questions about the nature of these bizarrely positioned worlds.

THE NATURE OF 51 PEGASI B

The majority of the discovered extrasolar planets do not seem to follow the trends observed in our own Solar System. 51 Pegasi, and many others such as Tau Boötis, 55 Cancri, Rho Coronae Borealis and the innermost planet in the Upsilon Andromedae system are located extremely close to their parent star (at around 0.05AU, with corresponding orbital periods of a few days!). The effective temperatures these worlds experience are consequently around 1400 K. This disturbs planetary scientists, how could such large planets (about 0.5–1 Jupiter masses) have possibly formed in such unusual places? (Lin *et al.* 1996).

The subsequent discovery of more of these bizarre worlds led astronomers to realize that they appear quite commonplace, although their initial discovery is a natural product of the detection method. These planets have a greater gravitational attraction to the star than a large planet further away, and hence would be discovered first. As detection technology improves, and the lengths of continued observation increase, it is expected that these so-called 'hot Jupiter' planets are common in recently discovered systems, but are likely to be quite rare in the Galaxy as a whole.

MIGRATION THEORY

The innermost planets of any given system were initially expected to consist of heavy silicate materials, like the four inner terrestrial worlds of the Solar System. Mercury, in particular, is thought to hold a substantial iron core, perhaps the remnants of heavy material which was primordially present very close to the Sun. To find a low-density gas giant so close to the parent star disturbs theorists. Did it originally form there? The currently 'accepted' theory to explain this involves the migration of the giant planets towards the star, as their orbital resonance locations change with time (Lin *et al.* 1996). Just why this has happened to the Upsilon Andromedae system and to others, while not apparently to our own, is a mystery. This makes us appreciate the stability of our own system.

SYSTEM STABILITY

The stability of the system is of the upmost importance when dealing with the prospects for life there. The system must be stable over the 100-million-year period thought necessary for the appearance of the first lowly lifeforms. This migrational history would have a direct bearing on the presence of small Earth-like planets. It is likely that any terrestrial mass world would be found in orbit of the giant planets, after being swept up. That is if they were not ejected from the system entirely. It is around these giant planets therefore, that we should look with future dedicated space-based missions for the classical signatures of life. An 'Earth' located around the innermost planet would not be a pleasant place to live!

THE UPSILON ANDROMEDAE SYSTEM

As far as we know, the Upsilon Andromedae system is unique, in consisting of three confirmed planetary bodies. This discovery raised a great deal of media interest at the time, as the first 'extra-solar system'. Upsilon Andromedae is well placed for observation from mid-northern latitudes, and consequently was the target of a photometric study which was undertaken as a final-year undergraduate project at the University of Hertfordshire Observatory. (Weldrake, 2000). The results focused on determining the conditions of the planets, the location of the habitable zone of the star (the region where the temperature is just right for an Earth-like planet to exist), and discussing the results to see whether this system is suitable for the development of life. It turns out that this system may not be as inhospitable for life as we might originally have thought. The results from these analyses are to be presented here.

Upsilon Andromedae was initially reported in 1997 to have a Doppler velocity perturbation, which was consistent with an approximately Jupiter-mass body orbiting at a distance of only 0.059AU (Butler *et al.* 1997, Butler *et al.* 1999), another apparent 51 Pegasi-type world. Follow-up observations from both the Lick Observatory and AFOE (Advanced Fibre Optic Echelle) Planet Survey Programs confirmed this periodicity, and found that simple models accounting for a single body orbiting the star could not totally account for the observations (Butler *et al.* 1997). Further analysis indicated that the surplus variations could be accounted for by additional bodies orbiting with periods of 242 and 1,269 days, respectively. This implies the presence of two additional planetary companions with masses corresponding to 2.0 and 4.1 Jupiter masses respectively (Butler *et al.* 1999).

The possibility of three about Jupiter-mass planets all orbiting within 2.5AU of each other immediately raises questions about the dynamical stability of this system. Subsequent computer modelling of the system (Laughlin & Adams, 1999) has shown that such orbits are stable over the main-sequence lifetime of the parent star. Although the innermost planet has a small eccentricity (a measure of the ellipticity of the orbit), the two outer planets seem to follow the observed trend, that outer planets have eccentricities of greater than 0.1 (Butler *et al.* 1999).

Table 1. The orbital parameters for the Upsilon Andromedae system

	Upsilon Andromedae 'B'	**Upsilon Andromedae 'C'**	**Upsilon Andromedae 'D'**
Orbital period	4.6 days	241.5 days	1295 days
Eccentricity	0.03	0.30	0.29
Orbital radius	0.059AU	0.83AU	2.50AU
Mass (Jupiters)	0.71 Jup	1.98 Jup	4.11 Jup

STELLAR DATA

Upsilon Andromedae itself is suitable for a study into the prospects for life. It is a single main-sequence star of spectral type F8, and is consequently slightly hotter than the Sun, with a surface temperature of 6100 K, compared to the Sun's photospheric temperature of 5770 K. It is located fairly nearby, at a distance of about 14pc (45 light-years) and consequently has a luminosity of some three Suns.

The following is a list of the apparent magnitudes and stellar parameters determined from the photometric analysis of Upsilon Andromedae at the University of Hertfordshire Observatory.

Table 2. The apparent magnitude of Upsilon Andromedae in V, B, I & R filters

Band	**Apparent Magnitude**
Visual (550nm)	+4.06±0.01
Blue (440nm)	+4.59±0.01
Infra-red (800nm)	+3.10±0.02
Red (700nm)	+3.61±0.01

Table 3. The computed parameters for Upsilon Andromedae

Parameter	**Value obtained**
B–V	0.53±0.02
C	0.42±0.02

Table 3. (cont'd)

V–I	0.96±0.03
I–R	−0.51±0.03
Temperature	6080±80 K
Spec.type	F8V
Colour factor, Q	−0.36±0.01
(B–V)o	−0.21±0.01
E(B–V)	0.65±0.02
E(U–B)	0.47±0.02
U–B	0.05±0.01
U	4.64±0.02
(U–B)o	−0.42±0.03
Luminosity	3.8±1.5 L(sun)
Mass	1.4±0.2 M(sun)
Radius	1.7±0.4 R(sun)
Abs.Mag	+3.35±0.43
Distance	13.9±0.4 pc
Dust Absorption	~0.01 (neg).

PLANETARY SYSTEM ANALYSIS

From the above photometric data, it is possible to investigate the conditions of the orbiting planets. Firstly, from the stellar luminosity, it is possible to determine the effective temperatures of the worlds. We can do this by considering the amount of heat the planet receives from the star at its orbital distance, and draw immediate conclusions as to the prospects for life there. Also, an estimate of the location for Upsilon Andromedae's habitable zone can be made, by considering the distance any planet would need to be from that star to have the same effective temperature as the Earth (255 K). The question of whether the habitable zone is stable must then be answered, to see if it is at all likely a planet can exist in this position.

PLANETARY EFFECTIVE TEMPERATURES

Using the planetary orbital semi-major axis values, (Butler *et al.* 1999) and the luminosity of Upsilon Andromedae, the following effective temperatures were obtained for the three planets in this system. Values were found for both the periastron (closest point to the star in the orbit), and the apoastron (farthest point from the star). This is necessary when the planetary orbit is non-circular, as the planet would experience different temperatures at different points in its orbit.

Table 4. The effective temperatures of Upsilon Andromedae planets

	Planet 'A'	**Planet 'B'**	**Planet 'C'**
Periastron	1426±146 K	426±44 K	271±28 K
Apoastron	1372±141 K	337±35 K	186±19 K
Average	1399±144 K	382±40 K	229±24 K

As can be seen, the innermost 51 Pegasi-type planet 'A' experiences extreme temperatures, as expected from its proximity to the host star. This temperature (some 1400 K, 1100°C), when first calculated caused some speculation as to the nature of the planet itself. At such high temperatures, above the melting point of many metals, it was initially expected that a massive Jupiter-type gas giant would lose its atmosphere quickly to stellar radiation pressures. If that was the case, then it conjures up attractive ideas of the planetary atmosphere gradually being bled from the planet, and forming an expanding gas shell moving outwards through the system. Such notions, however, seem unlikely, if we consider the mass of the planet and the recent results of the transit of HD209458. The density value derived directly from the transit showed that the planet is a very large, extremely rarified gaseous body, much more rarefied than Saturn (which itself is less dense than water). This rules out the theory that the world is a large terrestrial body, which would perhaps sit better with the theorists. As the temperature of this innermost world is higher than the melting point of many silicate materials, it immediately tends to rule out a terrestrial nature for the place.

The next planet out, planet 'B' has a temperature which seems to hover around the 380 K mark, or some 100°C. At this temperature, a Jovian gas giant can exist quite comfortably. The eccentricity causes the temperature to vary somewhat, from around 150°C to 60°C. Such extremes would cause some changes to the planetary atmospherics, namely producing very high velocity winds. These are observed on the Solar System's gas giant worlds, and are expected to be reproduced universally, although on a larger scale depending on the temperature extremes.

The outermost planet, planet 'C', is seen to have a temperature hovering around the freezing point of water, but experiences quite a marked difference in temperature as it orbits the star, due to its relative high eccentricity. The extremes are a high of 0°C, to a low −90°C. Again, this extreme will cause high velocity winds, depending on the atmospheric makeup and thickness. Although these are more typical of Martian temperatures and conditions than terrestrial, we shall see that after further discussion, it seems that a small planet orbiting Planet 'C' is the best chance for the survival of lowly forms of life in the Upsilon Andromedae system.

HABITABLE ZONE AND STABILITY

By considering these temperatures, a location can be estimated for a planet to experience the same conditions as the Earth, the habitable zone, which is obviously very important when discussing the prospects for life in this system. When using the Earth's effective temperature of 255 K, a distance of some 1.3AU is derived. Such a distance seems plausible when considering the 3 solar luminosity value for the star. (Laughlin & Adams, 1999, have the habitable zone for Upsilon Andromedae centred on 1.7AU; however, the overall conclusion is valid).

Computer modelling shows that in this case, 1.3–1.7AU is a highly unstable place to be. The 4.1 Jupiter mass planet 'C' is quite nearby, which seriously disturbs the orbit of any hapless world located here. There are two scenarios for a terrestrial-mass world. Either the planet is perturbed so much by the giant that it is kicked out of the system entirely, or it will go into orbit around the giant. Given the likely migrational history of this system (Lin *et al.*, 1996), this capture hypothesis seems quite plausible. Therefore when considering the prospects for life here, only the hypothetical terrestrial mass worlds in orbit of the gas giants shall be considered.

CONDITIONS FOR LIFE

Considering these results, the outermost planet 'C' seems the most suitable for the promotion of life. However, there are other factors which we must consider, such as albedo, or the effectiveness of the planetary atmosphere itself in reflecting light and heat away, and the 'greenhouse effect'.

Planet 'B' is a little hot, with an average temperature of around 100°C. This would cause a terrestrial planet in orbit here to be extremely dry – a hypothetical desert planet. The atmosphere itself would have an important bearing on the conditions; we cannot just state the temperature and not consider the atmospherics of the place. The temperature variations caused by the ellipticity of the orbit can directly cause high velocity winds, as previously mentioned, and the rotation of the planet can be very effective at dissipating the heat on its night-side – especially if it is rotating rapidly. However, the subsolar temperatures would still be very high indeed.

If the planet has a high albedo, like Venus, a great deal of the incoming heat would be reflected away, before reaching the surface and adding to the planet's heat. This could conceivably lead to a more acceptable surface temperature. However, when dealing with high albedo worlds, the problem of the greenhouse effect arises. The atmosphere would likely be thick enough to trap the IR radiation from the star, leading to a cumulative increase in temperature. When considering such problems, planet 'B' is too hot for the general promotion of life. This leads to an attractive picture of a Venusian-style planet located here, orbiting the 2.1 Jupiter mass planet 'B'.

The immediate environment of planet 'C' is the most interesting in this system. The temperature here ranges from −90°C up to around 0°C at periastron, similar to Martian temperatures, although temperatures of this range do exist on Earth, and life is to be found there. The atmosphere would again play a very important role, perhaps more so than the case of planet 'B'. A greenhouse effect would considerably increase the prospects for life. When at apastron, the atmosphere would act effectively as a blanket, holding in the heat, and when located at periastron, it would act as a barrier reflecting most of the heat away. The thicker the atmosphere, the more effective is the result. Such an effect could cause an average temperature of around −40°C, not as inhospitable to life as the original results suggest.

FUTURE DETECTION

After discussing the chances of a terrestrial planet existing here and its likely conditions, what are the chances of actually detecting it with current and future technology? Not too bad it would seem. To comment on this, we must consider the wavelengths we must use to have the best chance of seeing the planets, along with calculations of the angular separations the planets are from the star as seen from the Earth.

The peak wavelengths at which each planet radiates can be easily found using Wien's Displacement Law, which relates peak wavelength to effective temperature. (This is assuming that the planets radiate as black-bodies – perfect emitters and perfect absorbers of heat.) The angular separations of planet and star can be found using trigonometry. By knowing the distance to the star, and by knowing the planetary semi-major axis (orbital distance), a value can easily be found. The results calculated from the previously displayed planetary temperature and stellar distance values are as follows:

Table 5. The peak wavelength and separation values for the Upsilon Andromedae system

	Planet 'A'	Planet 'B'	Planet 'C'
Peak wavelength (micrometres)	2.1±0.3μm	7.9±1.7μm	13.5±3.8μm
Angular separation (arc seconds)	0.0038±0.0002″	0.05±0.01″	0.16±0.06″

These values show that the regime at which we must look to visually 'see' these planets is set firmly in the infra-red part of the electromagnetic spectrum. This makes sense when we consider that we are dealing with radiation caused by reflected light from the star itself. The angular separation values show that the innermost planet is way too close to the star for direct imaging, at least at the present time, and planet 'B' is still a little close for comfort. Planet 'C' seems plausible, as the separation is well within the resolving power of the world's best instruments.

As we must look in the infra-red, the best choice at the present time with which to look seems to be the NICMOS camera on board the Hubble Space Telescope. This instrument has the right wavelength range and good resolution, but, before going ahead and making a proposal we must consider the problem of dynamical range.

Dynamical range is defined as the difference in brightness of one object to another. Upsilon Andromedae itself is magnitude +4, and a Jovian planet located 14 parsecs away (at the same distance as the star) is expected to have an apparent brightness of around magnitude +25. This is a factor of 250 million times fainter. In order to directly image the planet, our instrument must be able to operate over this range of brightnesses. It is doubtful whether the Hubble Space Telescope can perform this well, but it is certainly worth trying.

With regards to the near future, the DARWIN and Terrestrial Planet Finder (TPF) space-based missions are dedicated to this task, using a technique known as nulling interferometry to reduce the brightness of the star while increasing the light from the planet. It is expected that in a decade or so, after the launch of these revolutionary new instruments, the first direct images of an extrasolar planet shall be made, perhaps of Upsilon Andromedae 'C'.

The results presented here are interesting enough, but they are only scratching the surface of the information retrieved from photometric observations. As this work suggests, a great deal of scientific information can be gained from relatively simple observations. Many questions remain to be answered about extrasolar planets, questions regarding their origins and evolution, their physical and chemical makeup, and perhaps the most thought-provoking, the prospects for the harbouring of life. In the near future many of these problems shall be solved, and with the launch of dedicated instruments, many exciting opportunities exist for cutting-edge research in this field.

BIBLIOGRAPHY

Butler, R. P., Marcy, G. W., Williams, E., Hauser, H., Shirts, P. 1997, *Astrophysical Journal*, **474**, L115–L118.

Butler, R. P., Marcy, G. W., Fischer, D. A., Brown, T. M., Contos, A. R., Korzennik, S. G., Nisenson, P., Noyes, R. W. 1999, *Astrophysical Journal*, **526**, 916–27.

Laughlin, G., Adams, F. C. 1999, *Astrophysical Journal*, **526**, 881–9.

Lin, D. N. C., Bodenheimer, P., Richardson, D. C. 1996, *Nature*, **380**, 606.

Marcy, G. W., Butler, R. P., Williams, E., Bildsten, L., Graham, J. R., Ghez, A. M., Jernigan, J. G. 1997, *Astrophysical Journal*, **481**, 926–35.

Marcy, G. W., Butler, R. P. 1998, *Annual Reviews in Astronomy & Astrophysics*, **36**, 57–97.

Mayor, M., Queloz, D. 1995, *Nature*, **378**, 355.

Weldrake, D. T. F., 'The Extrasolar Planetary System of Upsilon Andromedae – A Photometric Study', 2000, held at University of Hertfordshire Observatory.

Wolszczan, A. 1998, *Reviews in Modern Physics*. Forthcoming.

Wolszczan, A., Frail, D. A. 1992, *Nature*, **335**, 145–47.

Dense Matter: A Hundred Years On

GEORGE COLE

It is a capital mistake, Watson, to theorize before one has data. Insensibly one begins to twist facts to suit theories instead of theories to suit facts.

SHERLOCK HOLMES

Our knowledge of the Universe has been transformed over the last 100 years. Then, the cosmos, our Galaxy, appeared small and the stars seemed more or less fixed in the sky, except for planets and comets. In 1784 the comet searcher Charles Messier listed 103 nebulosities in the sky that are not comets. What of dense matter? In 1783 John Michell asked, in a lecture to the Royal Society of London, whether it was possible to have a finite quantity of matter of sufficient mass that nothing, not even light, could escape from its surface. There could be no answer then, but we know now that this is possible if the speed of light is the ultimate speed. Such matter, if it exists, is said to form a black hole because it cannot radiate electromagnetic energy through its surface and so would appear as a black disc.

It turns out that the important thing is the ratio of mass to radius. If the mass and radius are respectively α and β times those of the Sun, calculations show that α/β must be greater than 200,000. It follows that the larger the mass the larger the size. As one example, for the Sun to become a black hole ($\alpha = 1$) the radius must be compressed to 3 km (1.9 miles). How this would be possible is not clear. Again, the Universe as a whole is another extreme. By definition it includes everything and allows nothing to escape from it. The present estimated mass of visible galaxies leads to a radius that would take light a minimum of

15 thousand million years to traverse. The 'Michell question' has been answered in the affirmative. The solution can have a wide variety of forms, some local and involving a very high mean density while others may be extensive with a very low mean density.

HUBBLE EXPANSION

Our worldview was transformed following the inauguration of the 100-inch astronomical telescope on Mount Wilson, California, in 1917. The dazzling new technology provided new knowledge especially in the skilled hands of Edwin Hubble. Within a few years our view of the Universe was transformed by three fundamental discoveries.

The first discovery resolved some of the Messier clouds into spiral structures made of stars – they are separate galaxies like our own. Long-exposure photographs of them showed that they contain Cepheid variable stars whose distances can be determined. These turned out to be millions of light-years away and not part of our Galaxy. The Universe is large, and our Galaxy is only a small part of it.

Secondly, studying the light from these galaxies Hubble found that (with the exception of the nearest galaxy in Andromeda) it is all shifted towards the red end of the spectrum. More than that, the shift is greater the greater the distance from us. This is the Hubble law, still valid today. The red shift is generally accepted as arising from a velocity Doppler effect showing that the galaxies are actually moving away from us with a speed that increases with the distance. In other words our Universe is expanding.

Finally, the expansion means that the Universe will be bigger tomorrow than today but was smaller yesterday. If it is possible to look backwards in time we turn the expansion round to a contraction. The volume becomes progressively smaller and the Universe more compact. Is it, then, meaningful to ask what might have happened long ago (how long ago we wonder?) when all the matter was contained in a very small volume? We can estimate a rough value for this 'age' if we suppose the current expansion rate to have been constant throughout. Presuming everything started from rest, the beginning might have been a few tens of billions of years ago.

THEORETICAL MODELS: ENTER GEOMETRY

These observations require interpretation in terms of a model if they are to be properly appreciated. This could be the previous Newtonian dynamics, with some ad hoc assumptions, but a more general model is Einstein's theory of General Relativity (GR). Surprisingly, the predictions of the two theories are often the same although the derivations and interpretations are different. These theories are well known so we need only make general comments here.

GR is a continuum theory of gravitation, involving accelerated motion. It generalizes the earlier Special Relativity Theory (SR) which was restricted to observers at rest or in uniform motion (the inertial observers). The coordinates used by the observers are a combination of space and time coordinates, already isolated as useful in the earlier classical theory. The initial development of SR was in terms of algebra but Minkowski (incorrectly) interpreted the differential equation for the propagation of a light beam in three-dimensional space as a light path in a four-dimensional pseudo-Euclidean flat space-time. This led to the introduction of the combination of space and time as a single geometrical entity, in contrast to Newtonian theory which treats three-dimensional Euclidean space independently of the time dimension.

GR replaces the flat pseudo-Euclidean geometry of SR by a curved space-time. Einstein chose a geometry of the type treated by Riemann in the 19th century, giving a finite but unbounded volume. It reduces to the Euclidean form in the limit of indefinitely great curvature. The curvature of the geometry is to represent the effects of the gravitational force. The elimination of the force of gravity provides a forceless mechanics and follows the early investigations in classical mechanics, especially those by Euler, Gauss and Hertz. The geometry is specified through its curvature. The path of a test mass in the presence of a gravitational field is represented by a special path (a geodesic path) in the chosen geometry. Actually, this is a beautiful generalization of the law of inertia discovered by Galileo for a Euclidian geometry in which the shortest path is a straight line. The theory is developed by making an expression for the known energy and momentum of the material source of gravity equivalent to the detailed curvature of the geometry. The equivalence requires the introduction of a constant of proportionality that involves the Newtonian constant of gravitation, because the new theory must be equivalent to the old one in the limit of very weak gravity.

How do we choose the geometry? It turns out that it can be specified only as far as an arbitrary constant. This arises because the theory must reflect the conservation of energy and of momentum of the material world. The constant may be zero or it may have a finite value, according to the conditions imposed on the system. It was called the cosmological constant by Einstein and has had an interesting history. The selection of the metric itself is an assumption. Einstein chose the simplest form because this will reduce the difficulty in a complicated theory. Some other choices have been investigated in a preliminary way since then, but have not led to any improvement in the predictions of the theory. The original metric has remained the choice in all applications of the theory so far.

The form of the curvature has been the subject of many observational studies over the years but very recent observations have suggested that space and time in the mean are flat. This implies that, while the mean curvature of space and of time is essentially zero (infinite radius of curvature), for very local regions the curvature may be substantial. Further data are awaited but the confirmation of this result has important consequences for the general application of the theory.

SPECIAL PREDICTIONS OF THE THEORY

A continuum theory allows the propagation of waves, in this case those of gravity. GR is a non-linear theory, which means that the simple superposition of waves is not possible. In this it differs from sound waves and light waves. The source will be any substantial movement of matter such as the orbiting of one massive star by another. Different sources will have different characteristic wavelengths, but in all cases the waves are predicted to move through space with the speed of light. None have been detected so far, but a number of laboratories are currently developing apparatus to detect them and positive results are expected over the next few years.

A non-Euclidean geometry has the feature that there are two 'straight lines', one for a material particle and another for light beams. Because the geometry is linked with gravity it follows that a gravitational field affects the trajectory of a light beam. Such an effect was confirmed by observations at the solar eclipse of 1919. More recently, 'ghost' galaxies have been observed surrounding some actual ones due to the bending of light round the galaxy. This has allowed the mass of the galaxy to be deduced together with information about the matter behind it.

Schwarzschild was first to treat a spherical non-rotating mass, M, using GR. His solution (the Schwarzschild external solution) shows the remarkable property that the radial coordinate does not go right to the origin of symmetry but stops with a spherical horizon at a distance $\beta(0)$ away. Inside, the geometry is entirely undefined and physics does not apply. The radius $\beta(0)$ satisfies the same ratio $\alpha/\beta(0)$ as in the first section for the Michell question. Kerr more recently found a comparable solution for a rotating spherical body (the Kerr solution). Solutions for an electrically charged Kerr body are also known.

An expression, proportional to the surface area, has been proposed to describe the entropy of such bodies. This contrasts with the entropy for normal matter, where the entropy is extensive and proportional to the volume.

SINGULAR REGIONS

Regions where the normal rules of physics break down are well known and are called singular regions. They are generally excluded and accepted as marking a breakdown of a theory that must be generalized to eliminate them. The result of achieving this is a deeper knowledge of the alien region. The stable objects we observe in the Universe and describe in theoretical models involve a balance between two influences (or forces) and gravity is often one of them. Examples of this: a planet involves a balance between gravity and the strength of constituent atoms; a main sequence star involves the balance between gravity and the thermal energy (kinetic energy) of the constituent atoms; a white dwarf star involves the balance between gravity and the resistance of electrons to compression, called the quantum degeneracy force; a neutron star involves the balance between gravity and the resistance of neutrons to oppose compression, another application of degeneracy. The density is of nuclear magnitude and the escape speed is very close to that of light. Finally, gas clouds of low density are themselves in an equilibrium between the kinetic motion of the constituent atoms and the gravitational attraction between them. If a density fluctuation within the cloud gives a small volume object a larger gravitational pull than its neighbours, matter can condense to form stars.

If there are difficulties in finding a partner to form a balance in a particular case, it is because we have incomplete physical knowledge. For instance, a very massive star maintains a balance between thermal forces and gravity. It will collapse when its thermonuclear fuel is used up but may have a gravitational energy far in excess of that appropriate to form a neutron star. With no known resisting force of sufficient strength, it must collapse into the unknown state below the corresponding Schwarzschild horizon becoming a black hole. New knowledge could change this interpretation. Results in high-energy physics suggest that the neutron is a composite particle and can be broken up into the constituent quarks. These could provide degenerate energy to resist gravity. There are indications that even quarks have an internal structure which may also be available to resist gravity. This could lead to equilibrium forms of matter even more dense than neutron stars with escape speeds approaching that of light. Observation could decide: matter will spiral into a black hole with micro-second rotation periods, but could be simply absorbed by a collection of masses instead.

GENERAL EVOLUTIONS

It is well known that GR and Newtonian theory share a common set of predictions about the evolution of universes under gravity. The galaxies are treated as points in motion with kinetic energy K in a continuum with gravitational energy E. There are three direct equivalences between Newtonian ideas and GR:

1. K greater than E $\Longleftrightarrow$ positive continuum curvature
2. E greater than K $\Longleftrightarrow$ negative continuum curvature
3. E equals K $\Longleftrightarrow$ zero curvature (flat space)

The dynamical behaviour depends on whether the particles are allowed to move under gravity or whether they are constrained by gravity. This depends on the quantity of matter, that is the mean matter density. The results are:

1. The material expands continually from the origin 'for ever' (a singular solution).
2. The material expands at first but is pulled back later by gravity, thereafter oscillating between a maximum radius and the origin.
3. The material expands continually from the origin but comes to rest after an indefinitely long time (another singular solution). This is the critical solution.

These solutions apply whether the cosmological constant is zero or has a finite value but with slightly modified time scales. The motions in 2. can be rather complicated for special values of the constant.

A REAL EXPANSION? THE MICROWAVE BACKGROUND

Einstein initially developed a static Universe by defining the cosmological constant appropriately, but the Abbé Lemaître first recognized the dynamic solutions of the previous section, before Hubble's discoveries. The combination of theory and observation led to an expansion model in which an initially compact collection of material at very high density began a continuous expansion. There was discussion at one time about whether the initial condition was hot or cold but high compression was taken to mean a high temperature and so a Hot Big Bang. The beginning is to be a singularity where the material density and the temperature are each indefinitely great. Material will be entirely in its most fundamental form of radiation. This origin will also produce the space to expand into and will be the origin of time. Is there other evidence for such an extraordinary state?

Evidence came in 1963 with the discovery of a uniform background radiation at microwave frequencies coming from all regions of the sky. It was found to have a full black-body distribution of frequencies corresponding to a temperature of 2.75 K. More recently the radiation has been studied by satellite observations in the COBE mission, and variations of temperature of one part in 10,000 have been identified and plotted. It is now accepted that the radiation is a remnant of the initial fire ball of the Hot Big Bang. Encompassing the whole Universe, the expanding material cannot have lost heat energy through its boundaries (the expansion is adiabatic) and the temperature will have fallen consistently as a result. The models of the previous section can be applied equally to matter or to radiation so the expansion profile can be predicted if the theory is valid to that period. It is, of course, an assumption that GR applies to such situations.

There are two approaches to the expansion, one running forwards from the presumed initial epoch and the other moving backwards from now. The former is the standard approach based on knowledge from high-energy physics. The temperature falls during the expansion allowing elementary particles to materialize out of the underlying vacuum. The main action takes place during the first three to four minutes. Hydrogen, helium and some lighter atoms are formed and molecules begin to appear after some 500,000 years. At this stage matter and radiation decouple, the radiation ultimately cooling to become the observed microwave background.

The arguments are not complete, largely because our knowledge of the underlying physics is not complete. Our present knowledge is gained from high-energy experiments and these involve equivalent temperatures no greater than 10^{14} K. This would be appropriate to very early stages of the expansion but is not sufficient to provide the wide array of fundamental particles necessary to give a quantitative picture of the processes very near to zero time. The alternative approach, therefore, is to move backwards until the current limits of experimental knowledge are reached. This has not been done in detail so far but would not yet allow the first four minutes to be properly accounted for.

POINTS TO RESOLVE

The expansion theory has its difficulties quite separately from lack of early detail. These are centred on the link between the Universe now and how it was initially. The early fireball would be very small and highly turbulent. The speed of expansion at first would be substantial. The different regions could not have been in communication with each other and this makes it difficult for the Universe to achieve its present isotropy and homogeneity (the so-called smoothness problem). This led to the rather bizarre suggestion that different regions expanded into separate universes each showing the required homogeneity and isotropy. There seems to be no possibility of observing the validity of this assertion.

One suggestion, very popular in recent years, is the assumption of an inflation of the Universe to many thousands of times its initial size, very soon after the start of the expansion. This is the inflation theory describing the release of energy in a phase change to drive the expansion. This approach is, however, not valid. The equations of GR are known to be adiabatic, but the inflation is incompatible with this because energy is to be released. The recognition that space is mainly flat means that inflation must have the different role of accounting for the gross curvature of early times. In all these thoughts it is still not clear when and how galaxies formed.

An alternative suggestion, made some time ago, is that the speed of light was greater in the earliest periods than now. This is, of course, contrary to GR. If the early Universe was very hot, it is tempting to wonder whether the light speed might increase with temperature. Light travels in the vacuum underlying ordinary matter. The vacuum is a quantum concept recognized by Dirac in the 1930s. He predicted antiparticles in an underlying quantum world associated with the mutual transitions between subatomic particles and energy, in and out of the vacuum. Casimir much later demonstrated that the vacuum is, indeed, the source of energy. Details of the vacuum are still not known.

INVISIBLE MATTER

The quantity of observed matter has been found insufficient to account for the dynamical behaviour of the Universe for two reasons. First, the stability of rotating galaxies requires more matter than is observed. Second, galaxies occur in clusters, some (such as in the constellation Virgo), containing several thousands of members. The long-term stability of such clusters requires more matter than seems to be there. It turns out that less than 10 per cent of matter can be seen. The missing matter is called 'dark matter' and has a gravitational influence even though it is hidden from sight. This new assessment of mass still leaves the gravitational energy less than the kinetic energy. Very recent observations have shown that space is flat in the mean, so the geometrical curvature is infinite. We have seen earlier that if K and E are equal then the material density has the critical value. It seems the mean density must be supplemented by yet more matter which cannot be seen.

More is yet to come. Very recently it has been discovered that the

Hubble expansion is accelerating and not decelerating as was expected. This implies yet another source of energy to drive the expansion, now called the 'dark' energy. This is presumably centred in the quantum vacuum underlying the macroscopic world, but the source of it is quite unknown. Presumably the energy carries a mass and this mass must be sufficient to lift the mean density up to the critical value.

EPILOGUE

It is interesting to notice that to this point the one definite feature of the Universe has not yet been mentioned – the presence of life, on Earth at least. We might seek life elsewhere, but life is definitely here. The occurrence of life implies certain constraints on the values of the fundamental constants of nature. This has often been interpreted as meaning that there can be Universes without living materials but the only experience we have shows this not to be the case. We must surely be a deciding factor in the structure of the Universe.

So what of the next hundred years? We can predict that improving technology will surely provide ever more precise data of what we know and will reveal features of the Universe as yet unsuspected. The generalization of GR and a deeper understanding of the quantum vacuum may allow us to determine the internal structure of black holes and set them in the cosmic hierarchy. We can expect to detect events further back in time, nearer to the initial big bang, and learn more of the energetic Universe at the earliest times. Perhaps the gravity waves from even earlier times will be detected and studied. A pressing problem is the discovery of the role of living materials in the Universe. Is life an evolutionary feature of the cosmos? – are the animate and inanimate worlds mutually linked? Who knows what new discoveries lie round the corner? The future promises to be very exciting.

Through A Glass Clearly

CHRIS KITCHIN

When Galileo first looked at Saturn, he thought he saw a triple planet. The optical quality of those early telescopes was so poor that even major features like the rings of Saturn were blurred out of existence. Since Galileo's time though, we have come to understand thoroughly the optics of telescopes. So today even an amateur's small telescope not only shows Saturn as a planet surrounded by rings, but reveals details of the rings such as Cassini's and Encke's divisions.

The sharpness of an image through a telescope is determined by its angular resolution. For a small telescope this is usually around one second of arc (1″) or $^1/_{3600}$ of a degree. The unaided eye has a resolution of around three to five minutes of arc (3–5″), so even a small telescope will improve the sharpness of our view by a factor of 200 or more.

Our knowledge of optics tells us that as telescopes get larger in diameter the sharpness of the images that they produce should improve. The sizes of the angles that can be resolved should thus decrease as the telescopes get bigger. A 12-cm (5-inch) telescope has a resolution of 1″, a 1-m (40-inch) telescope a resolution of 0.12″, a 2-m telescope a resolution of 0.06″ and so on (see Figure 1).

With today's 8-m and 10-m telescopes we should therefore be able to see details in Saturn's rings as small as 100 km (60 miles) across. Yet when such telescopes are used to observe Saturn, or any other object, they rarely show details any finer than those to be seen through an amateur's 12-cm (5-inch) telescope.

This blurring results not from poor quality optics, as it did in Galileo's time, but from the effect of the turbulent atmosphere of the Earth. We peer out at the Universe through a thick layer of gases that are constantly moving in all directions and varying in density and transparency. We can see the effects of the atmosphere even without a telescope through the twinkling of stars. The blurring effect of the atmosphere means that from most observing sites the resolution of a telescope, whatever its size, is limited to about 1″, which is the theoretical resolution of a 12-cm telescope.

This is not quite the whole story, however. Because the atmosphere is constantly in motion, its effects upon an image are constantly changing. Sometimes therefore its effects will blur the image to more than 1″, but sometimes also the atmosphere will stabilize briefly so that resolutions of better than 1″ may be reached. Experienced observers can thus often see far more details than seem apparent to the casual observer, because they have learnt to make use of those brief intervals of clarity. In this way telescopes larger than 20 cm do occasionally reach their theoretical resolutions (see Figure 2). However even from the best sites in the world, telescopes of 1 m or more in diameter will never have a stable enough atmosphere to reach their limiting resolutions.

The atmosphere rapidly thins out at higher altitudes. At the top of Everest (altitude 8.8 km, or 5.5 miles) you would be above two-thirds

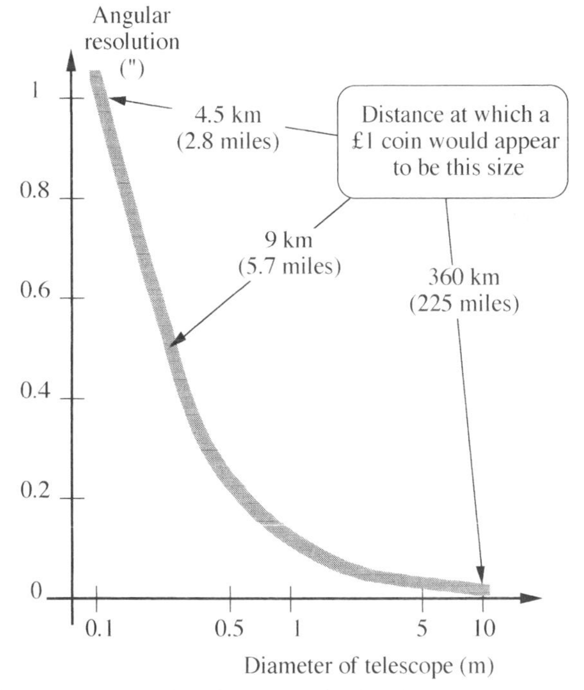

Figure 1. The angular resolution of telescopes of various sizes.

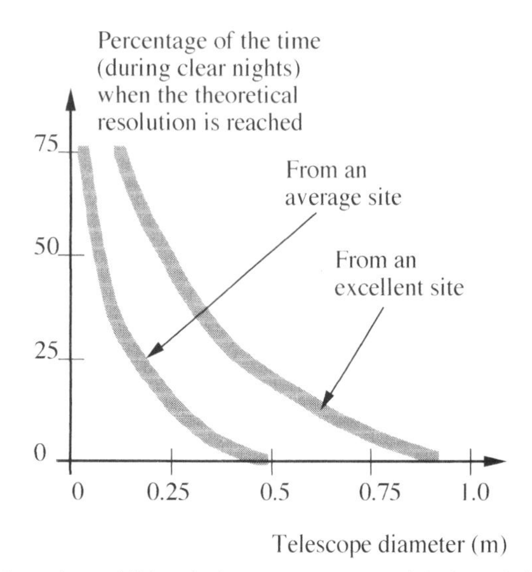

Figure 2. Approximate abilities of telescopes to operate at their theoretical resolutions.

of the atmosphere, and at a height of 20 km (12.5 miles) only 1 per cent of the atmosphere is left. Furthermore most of the atmospheric turbulence is in the lower layers of the atmosphere. One way of reducing atmospheric blurring is therefore to move the telescope to as high an altitude as possible. To this end most large professional telescopes are now sited on the tops of mountains in Hawaii, the Canary Islands, the Andes and elsewhere. Even greater heights can be reached by putting the telescope aboard an aeroplane, or suspending it from a high altitude balloon.

Ultimately, of course, the effect of the atmosphere can be completely eliminated by launching the telescope into space. But this is a very expensive option. The Hubble Space telescope for example has cost around £1000 million ($1,600 million) so far. This would suffice to purchase 50–100 ground-based telescopes of the same size, or up to 20 10 m instruments such as the Keck telescopes.

While there are plans to put telescopes larger than the HST into space, the Next Generation Space Telescope (NGST) for example is hoped to be 6–8 m in diameter, it would be of enormous benefit if the existing large ground-based telescopes could be used to give images as sharp as they ought.

One way of enabling large telescopes to produce sharp images is similar to that used by all of us when we take holiday snaps of moving targets; if an exposure of $^1/_{30}$ second gives a blurred image, then we shorten the exposure to $^1/_{60}$, $^1/_{120}$, $^1/_{250}$ second etc., until we manage to freeze the motion. Now the atmosphere is changing on a timescale of a few milliseconds. So in order to freeze its effects, an exposure of $^1/_{1000}$ second or so is needed. Since imaging astronomical objects usually requires exposures of many seconds, or minutes or even hours, a large telescope is needed with very sensitive detectors to show anything at all in a millisecond. Nonetheless it is possible to take very short exposures of the brighter objects, and so to freeze the turbulent motion of the atmosphere. When such an image of a star is obtained it is not, however, as might be expected; instead of the single sharp image at the resolution limit of the telescope, many sub images are seen spread over an area of 1 second of arc or so. (Figure 3).

The origin of these sub-images, or speckles, lies in the way that

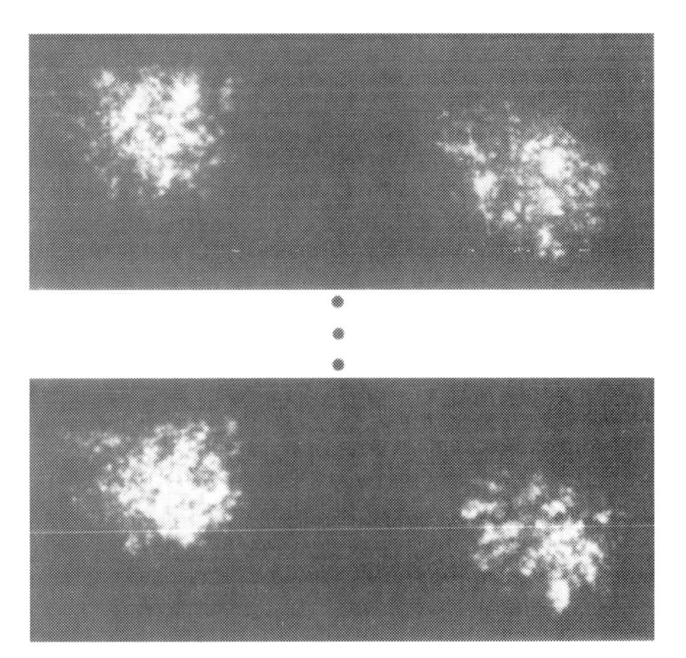

Figure 3. Short-exposure direct image of Zeta Cancri taken using the 3.6 m telescope at the European Southern Observatory. [Courtesy of the European Southern Observatory]

atmospheric turbulence affects the light coming from the star. The light spreads out from the star in all directions, and so forms a wavefront which is spherical and centred on the star. But the distance to the star is so great that the curved wave front is effectively flat when it reaches the top of the Earth's atmosphere. That flat wavefront is then chopped by the unevenness of the atmosphere into small segments, a few tens of centimetres across, which travel unevenly. Some move very slightly faster or slower than their neighbours. The difference in velocities is tiny, but sufficient that the segments of the wavefront arrive at the surface of the Earth (or the main mirror of the telescope) delayed or advanced by a few fractions or small multiples of the wavelength of light. (Figure 4).

When this broken-up wavefront is brought into a focus by the telescope, the delays and advances induced by the atmosphere lead to some parts of the wave being out of step (or phase) with others. These out-of-step segments then produce an interference pattern similar to that which occurs when light passes through two slits (Figure 5). Each segment will interfere with every other segment, but generally only the central, bright fringe will be visible as a speckle in the frozen image. Even so, since the segments are typically 15–30 cm (6–12 inches) in size, a 1 m diameter telescope will have 20 to 30 segments of the wavefront hitting it at any given instant, and so thousands of resulting speckles.

The interference patterns can be converted back to a normal image by optical processing in the laboratory. Usually the results of many images such as that shown in Figure 3 need to be combined, but an image at close to the theoretical resolution of the telescope can then sometimes be obtained (Figure 6).

High-resolution images obtained from interference speckles, however, can be obtained only for the brighter objects which have a comparison star nearby, and so the process is limited in its application. In order to get sharp images of the majority of objects in the sky, we need to try to remove the effect of the atmospheric turbulence. This is now being undertaken by using adaptive optics.

The aim of adaptive optics is to restore the broken-up wavefront from the object to its original uniform structure. To do this, the segments of the broken-up wavefront must be delayed or advanced until they are again all in phase. This can be accomplished by reflecting the wavefront from a mirror whose sur-face is the same shape as the wavefront (see Figure 7), but with half the vertical distances between

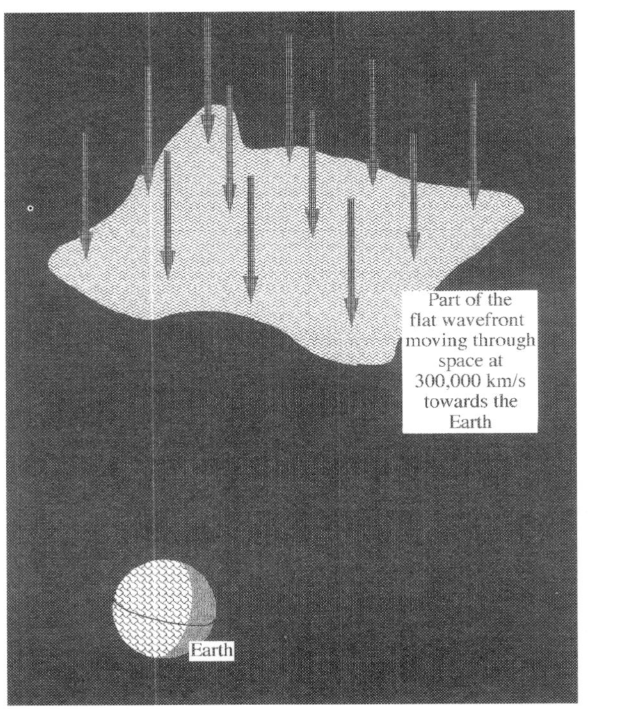

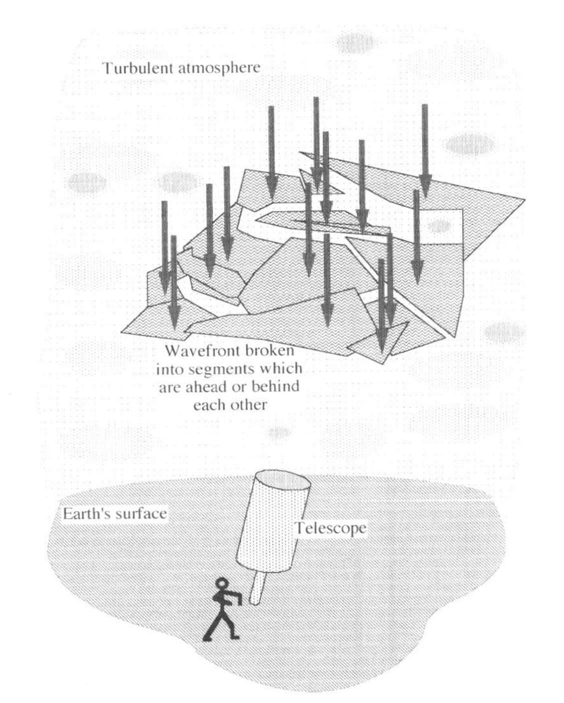

Figure 4. The effect of the Earth's atmosphere on an incoming flat wavefront from a distant object.

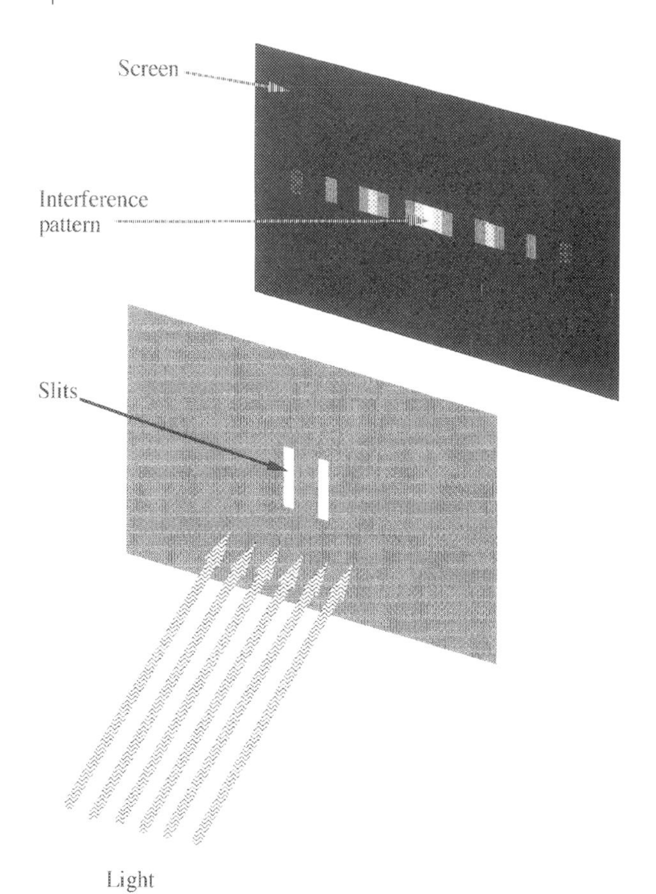

Figure 5. The interference pattern produced when a beam of light passes through two slits.

the mirror segments compared with those of the wave front segments (because the leading segments travel down to 'their' bit of the mirror *and* back again, while the trailing segments do not).

Now if the distortion due to the atmosphere were constant, then a simple mirror could be shaped to provide the restored wave front. But, as we have seen in discussing speckle imaging, the distortions caused by the Earth's atmosphere change every few milliseconds. The mirror must therefore also change its shape to match the changing broken-up wave front. An adaptive optics system thus requires a mirror whose shape can be changed quickly. In addition there must be a system for detecting the distortions in the wave front, in order to determine the shape required for the mirror.

The main system used to detect the distortions in the wave front is known as the Hartmann sensor, although other systems such as the shearing interferometer are also employed. With the Hartmann sensor a small proportion (typically 10 per cent) of the light from the telescope is diverted by a beam splitter through an array of small lenses (see Figure 8). The individual images produced by those lenses are then detected by a CCD detector. If there is no distortion, the images will be centred behind each lens. If there is wave front distortion, some or all of the images will move from their centred positions. The changes in positions of the images are then used to determine the wave front distortions.

A fully deformable mirror is produced from many small sections individually mounted on piezo-electric crystals (see Figure 9). But a big improvement in image quality can be obtained just by correcting the overall tilt introduced into the wave front by the atmosphere, and for this only a simple flat mirror is needed whose tilt can be altered. Tilt correctors are now available commercially at prices within the grasp of an amateur astronomer. The more complete correction given by a fully deformable mirror provides even better images (see Figure 10), but the cost is many times that of the tilt corrector.

Thus adaptive optics enables large telescopes to image stars at close to their theoretical resolution. Unfortunately since stars are point sources, the increased resolution provides us with very little more information. For adaptive optics to make a significant contribution, we need to be able to use it to image extended sources like planets, galaxies,

Figure 6. Direct image of Zeta Cancri obtained from speckle images (Fig 3) using the 3.6 m telescope at the European Southern Observatory. [Courtesy of the European Southern Observatory]

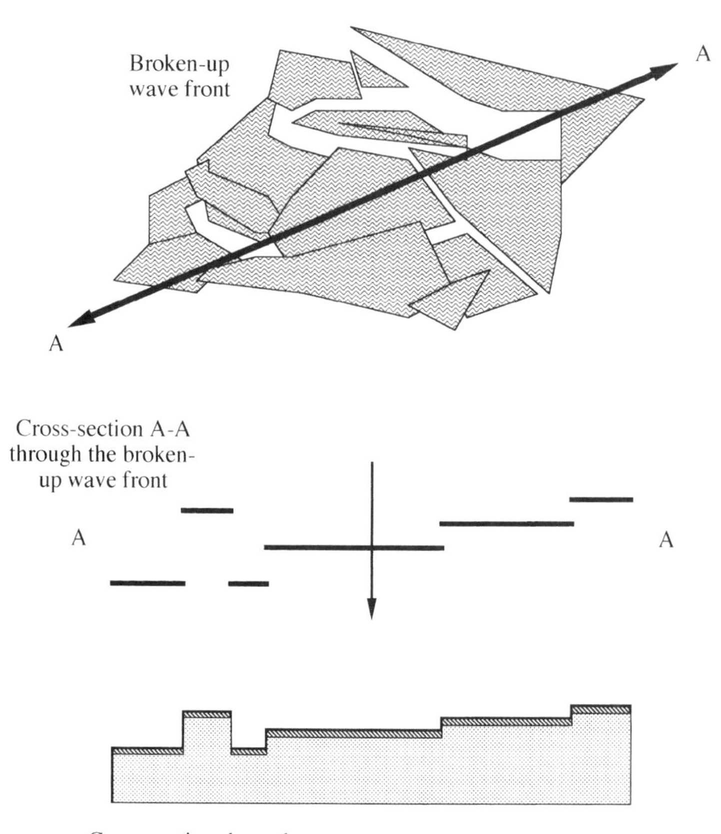

Figure 7. Restoration of the wavefront by reflection.

nebulae, star clusters etc. But the Hartmann sensor (and other similar devices) will work only for point sources. What is required therefore is a star close to the object that we are interested in, which can be used to generate the correction signal in order to sharpen the image of the extended object.

A comparison star used in this way must be close enough to the main object that its light is affected by the atmosphere in a similar manner (see Figure 11). The area of sky over which the atmosphere does have a uniform distortion is called the isoplanatic area, and is very small. Typically the isoplanatic area is no more than 20″ across; smaller than the size of Jupiter in the sky. The comparison star thus has to be very close indeed to the object that we are interested in or the adaptive optics correction will fail.

In practice therefore, few objects have suitable comparison stars available. To try to overcome this problem, there have recently been attempts to produce artificial stars. The artificial star is formed by shining a powerful narrow laser beam up into the sky. The laser is at the wavelength of one of the sodium D lines, and so causes sodium atoms high in the atmosphere to glow. Seen through the telescope the atoms shine like a faint yellow star. By moving the laser, the artificial star can be placed as close to the actual object that we are interested in as may be needed for both to be within the isoplanatic area.

There are still some problems with using an artificial star because

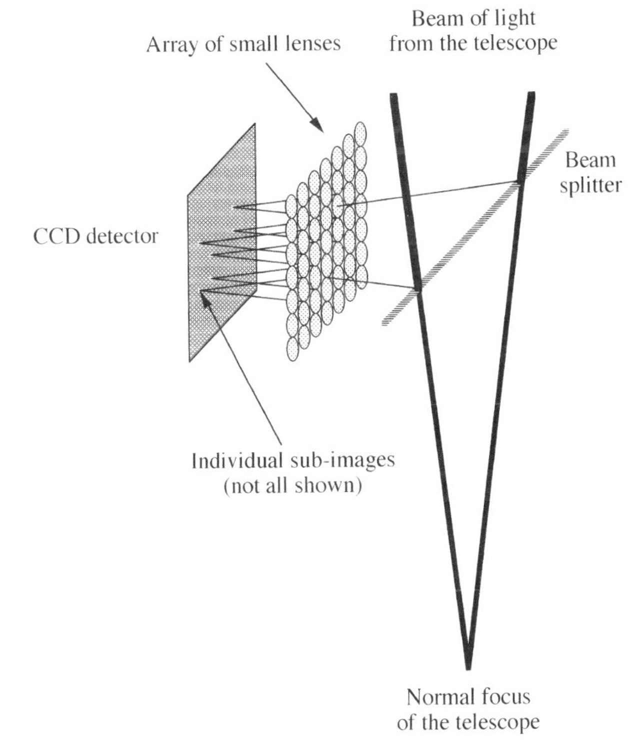

Figure 8. Detection of wavefront distortion using a Hartmann sensor.

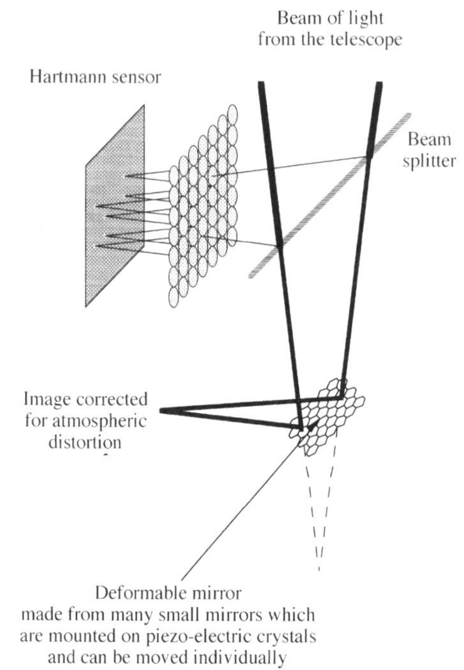

Figure 9. Correction of wavefront distortion.

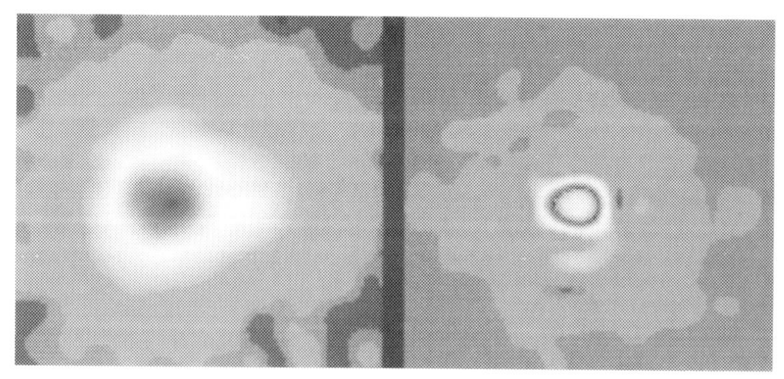

Figure 10. Images of Deneb (Alpha Cygni) at a wavelength of 2.2 microns taken with the 1.52 m telescope at the Observatoire de Haute Provence. The left-hand image is 'normal', the right-hand image is with the adaptive optics corrector in use. Deneb's diameter on the right-hand image is 0.37"; the theoretical resolution limit of the telescope in the infra-red. [Courtesy of the European Southern Observatory]

although the star is high in the atmosphere (about 90 km) (56 miles), it is not as far away as the object being observed. So the light paths through the atmosphere from the artificial star and the object in the sky are not quite identical. Nonetheless, with this last development, and nearly four centuries after Galileo, we may soon be able to 'see through a glass clearly' with even the largest Earth-bound telescopes.

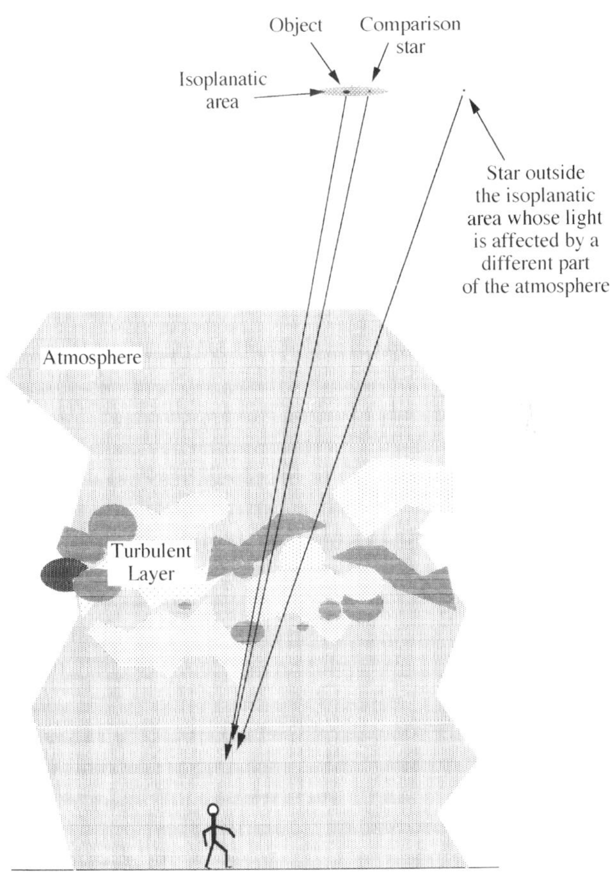

Figure 11. The isoplanatic area.

Astronomical Coincidence or Cosmic Design?

ROGER O'BRIEN

Well, I shall leave it to you to decide, but let us start with a coincidence that I know is one of Patrick Moore's favourites. When you are trying to imagine the scale of the Universe, take one inch (2.54 cm for the metric-minded among us) to be the average distance from the Earth to the Sun and the distance to the next nearest star will be 4.5 miles on the same scale. I think it is quite remarkable that the famous '1 inch to 1 mile' of the good old reliable Ordnance Survey maps is in very nearly the same ratio as between the radius of the Earth's orbit (which is one Astronomical Unit or AU) and a light-year. In numbers, 1 mile = 63,360 inches and 1 light-year = 63,290 AU. The Astronomical Unit is 150 million kilometres or 93 million miles and it takes light over eight minutes to cover that distance, making 1 AU equal to 500 light seconds. So, the diameter of the Earth's orbit is 1,000 light seconds or nearly 17 light minutes. I should not like to pretend that any of this is very important, but I find these odd things fascinating.

The most famous coincidence can often be seen. The Sun and the Moon are almost identical in size as we see them in the sky. This is the reason for the spectacular total solar eclipses of the kind which I (unfortunately) did not see from Penzance on August 11, 1999. The Moon's shadow only just reaches the Earth and traces a very narrow path. You have to be within that path to see a total solar eclipse. The Sun is roughly 400 times as big as the Moon and 400 times as far away from us.

Perhaps the most remarkable thing about this coincidence is that it is a temporary affair. Thanks to tidal friction, the Earth's rotation is gradually slowing down and the Moon is gradually receding from the Earth at about 4 cm (about 1.5 inches) per year. So, in the past, the Moon would have appeared larger than it does now and would have eclipsed the Sun more easily, but slightly less spectacularly. A noticeably larger Moon would also cover the chromosphere and some of the inner corona of the Sun and those prominences I had hoped to see in 1999. As the Moon recedes from us, its average angular diameter will get smaller and, eventually, it will not be able to cover the whole face of the Sun at all. Our descendants may see only annular eclipses, in which there is a ring of Sun visible round the dark disk of the Moon.

The diminishing of the Moon's size is not a rapid process and it will still appear at about two-thirds of its present size when the Sun starts to swell up and become a red giant. This is a very crude calculation, but it gives the right idea of the timescales involved.

There is another piece of serendipity of the same sort of order and that is the existence of Saturn's rings. Almost everyone, who has looked through a telescope at the most distant of the planets, known to the ancients, has been delighted by that wonderful sight. For myself, not only is the sight a great and enduring pleasure, but I enjoy showing Saturn to my pupils, especially those who have not seen it through a telescope before. The point at issue, here, is that those rings are almost certainly ephemeral on an astronomical timescale. In a few million years, solar radiation will begin to darken them. At the same time a subtler effect of the Sun's light will cause the orbits of the smaller particles in the rings to decay slowly and the planet will swallow them up. The rings will become thinner and darker as time passes. We are fortunate to be here, now, when the rings are mature enough to form their brilliant appearance, but have not decayed. The sky will be a poorer place without those magnificent rings.

Jupiter may not have a ring system to rival Saturn's, but it does have the Great Red Spot. At the time Voyager 2 reached it (August 24, 1989) Neptune also had a Great Spot, but it was dark rather than red. However, Neptune's spot and Jupiter's were both in the same southern latitude and have the same size relative to their planets. And both are anti-cyclones. Unfortunately for this coincidence, observations with the Hubble Space Telescope no longer show Neptune's Great Dark Spot – it has disappeared. If it is any consolation, the *El-Nino* phenomenon on Earth also seems to occur in roughly these same southern latitudes.

Let's stay talking about the sky for a moment or two. Are we grateful enough for the striking arrangement of the stars in the Plough that has two of them, Merak and Dubhe ('the pointers'), lined up very nicely with the even more fortunately placed Pole Star? There is nothing like this in the southern hemisphere. Of course, it is another temporary phenomenon. All these stars have their own proper motions, and over long periods of time, say 100,000 years, as these motions become obvious, the appearance of the constellations, which we take for granted, will change.

The set-up is ephemeral in another way, too. The Earth spins on its axis happily enough, but that axis describes a circle in the sky. At present, it points nearly enough for most practical purposes at the Pole Star. The movement of the axis will take the precise northern celestial pole first a little nearer to the star then past it and away from it. In less than 14,000 years (a mere twinkling of an eye, astronomically) the actual pole will be near the bright star, Vega. In another 14,000 years the celestial north pole will have gone the full circle and be back close to our Pole Star – I wonder if there will be humans to see it?

Back in the time of the Pharaohs, the north celestial pole was close to Thuban. This star is fainter than our Pole Star and less easy to find, even for people who lived under darker, less polluted skies than we. The coincidence does not end there. When Thuban was the pole star, Phad and Megrez were in line with it in much the same way as Merak and Dubhe are with our Pole Star. I think this needs a little emphasis: the only two stars that have been really worthy of the designation 'pole star' have had convenient pairs of pointers to confirm the identification. It is even more remarkable that all four 'pointers' are in the same part of the same constellation.

May 2000 saw dramatic coincidences. Apparently, all the ancient planets were in the same astrological 'House' on May 5. This highlights a difference between astronomy and astrology. The House or Sign does not coincide with the constellation of the same name and the whole lot were never in one constellation. Of course, the word 'planet' mean 'wanderer' (incidentally, I have always thought 'nomad' would be a better translation) and the ancient planets were: Sun, Moon, Mercury, Venus, Mars, Jupiter and Saturn. With so many objects concentrated in a band of sky 30° wide there were bound to be lots of conjunctions between pairs and triples. There is, of course, a catch. Notice that the Sun was involved, so most of these were daytime events and hidden in the bright-blue scattered light of the Sun or, for readers in Britain, probably obscured by cloud.

April 2000 saw a suitable prelude to all this. Saturn was moving slowly eastwards through Aries. Jupiter steadily overtook it. Mars came up still faster. April 6 had Mars close to Jupiter and the Moon only a few degrees away, nearer to the southern horizon. By April 16 Mars had overtaken Saturn and would lead the way into Taurus. All the while Jupiter edged up towards Saturn.

Writing this at the end of January 2001 I am conscious of the difficulties of describing events in my future to readers for whom it will be in the past. So, I shall ask my gentle readers, please, to excuse any confusion of tenses.

Following the month of May 2000 through on my computer, I see it open with Mars in Taurus and the Sun, Mercury, Venus, Jupiter and Saturn all in Aries (I am speaking of the constellation). As I look at the screen and move forward a day or two, the Moon enters from the right. By May 5 it is in Taurus and about to invade the Hyades – I rather like that picturesque phrase, which is a translation of a Chinese astronomical term. May 8 sees the Sun, Mercury, Jupiter and Saturn in a close huddle near the Aries/Taurus border. Mars and Venus form flank-guards, equally spaced east and west respectively. By May 16, the Sun has crossed into Taurus and Venus is crowded up with Jupiter and Saturn. About May 19, Mars and Mercury are close together and the other trio have not separated a lot. The line-up for May 30 is again striking: Mercury on the eastern border of Taurus; Jupiter and Saturn on the western; Venus, Sun and Mars strung out across the constellation from west to east. On June 1 the Moon rushes up from the west to join them, but Mercury, demonstrating its role as the swiftest planet, has moved well into Gemini.

Before we leave this parade of computerized planetary dancing, I must remark on one more thing. The summer solstice marks a traditional major turn in the year and, this time, the planets help it out. Venus and Mars are in conjunction, too.

For an astronomer, none of these conjunctions is more than a coincidence. The process is worth a look, if only to get some idea of how puzzling the motions of the planets were to most of our ancestors.

One of the more surprising coincidences is between the sizes of Mercury and Callisto. Their diameters are 4878 km (3031 miles) and 4806 km (2986 miles) respectively. The discrepancy is less than 2 per cent. As usual, the similarities do not end there. Both bodies are heavily cratered and lack marked changes in surface colour and brightness. Both have a major impact basin dominating one hemisphere – the Caloris Basin on Mercury and the Valhalla Basin on Callisto.

Mercury and our Moon also look pretty similar, especially to a non-expert eye (e.g. mine), but the differences in size, density, etc. are considerable. Any real expert can tell them apart. If we look farther afield, we can find two bodies that are very similar to our Moon in size: the Jovian moons, Io and Europa. Neither of these is in the least similar in appearance to our Moon.

Planets do come in pairs in all sorts of ways, for example, Earth and Venus are much the same size, so are Uranus and Neptune, and, stretching the point a bit, so are Jupiter and Saturn. However, none of these coincidences is quite up to the Mercury-Callisto standard.

Earth and Mars have similar length days, so do Jupiter and Saturn and Uranus and Neptune.

Mars and Earth currently both have winter perihelia so their southern hemispheres have hotter summers and colder winters. The combination of the summer Sun being higher in the sky for longer each day and nearer, as well, boosts the summer temperatures. In winter, the opposite is true: the Sun is lower in the sky for a shorter time and farther away. This is much more significant for the smaller colder Mars, with its rather elliptical orbit, than for the Earth, which has a more nearly circular orbit. I really cannot say whether any of this is significant, but it caught my eye.

Quirks of dynamics mean that Venus tends to have the same face towards the Earth at each inferior conjunction – not that one would know by ordinary observation. The early attempts at radar mapping, working from Earth, gave results that seemed anomalous, because they were undertaken around the time of inferior conjunction when Venus was near the Earth, so the returning radar signal bouncing off the planet's surface would be at its strongest.

I also like the coincidences around the length of our month. The rate of rotation of the sunspot belt we observe on the Sun is about 28 to 29 days. Mercury's year is 29.3 of our days longer than the time it takes to spin once on its axis. So Mercury takes two of our months to spin once on its axis and its year is three months. The lunar synodic month is 29.6 days (full Moon to full Moon).

The Moon is high in late summer and that provides light for northern hemisphere harvest gatherers to continue working by. This is the famous Harvest Moon. Is this a coincidence?

This brings me to resonances, which are not coincidences, but have a similar numerological fascination. Mercury's orbital and rotational periods are in the ratio 3:2. Our own Moon's orbital and rotational periods are the same, so the ratio is 1:1. Pluto and its moon, Charon, are even more tightly linked up: both Pluto and Charon share the same rotational period and Charon's orbital period is also the same so that is 1:1:1.

In the outer Solar System, Jupiter's and Saturn's orbital periods are very close to the ratio 2:5. Beyond that, the ratios of the orbital periods of Uranus, Neptune and Pluto are related to Jupiter's by multiples of 7:

Table 1.

Planet	**Orbital period (years)**	**Ratio to Jupiter's orbital period**
Jupiter	11.86	1.0000
Saturn	29.46	2.4840
Uranus	84.01	7.0835
Neptune	164.79	13.8946
Pluto	248.54	20.9562

In the Jovian system, the orbits of Io, Europa and Ganymede are dynamically linked. Their orbital periods are in the ratios 1:2:4. Callisto doesn't fit the sequence, but perhaps it is too busy looking like Mercury.

In the Asteroid Belt, there are gaps where you do not find asteroids. These are called Kirkwood Gaps after the man who identified them. One gap corresponds to a body having an orbital period exactly half that of Jupiter. Every second of that body's years, it passes between Jupiter and the Sun. Jupiter's gravity perturbs the orbit until it is no longer so exactly lined up. Similar simple relationships between the orbital periods define the other gaps.

The Cassini division in Saturn's rings is related in the same way to the moon, Mimas. Anything orbiting in the Cassini division has a period half that of Mimas.

Now, what about the famous Bode's Law? Can we really model the radii of planetary orbits with a fairly simple mathematical series? Nowadays, it would be a bold astronomer, who said Bode's formula was a real law, but it once had quite a good pedigree. By the way, the formula seems to have been devised first by Titius, but Bode seems to have handled the publicity with greater skill and got the law named after him. Let's have a look at it:

You start with a simple series: 1, 2, 4, 8, 16. If you then multiply each number by three, you get: 3, 6, 12, 24, 48. Finally you add four to each and create a new first number, 4, to complete the series. Tabulate these against the measured planetary distances and this is what you get: (I have given the average radius of each planet's orbit in 'Bodes' where 10 Bodes = 1 AU).

Table 2.

Stage 1	Stage 2	Bode's Law	Orbital radius	Planet
		4	3.87	Mercury
1	3	7	7.23	Venus
2	6	10	10.00	Earth
4	12	16	15.24	Mars
8	24	28	27.70	Ceres*
16	48	52	52.03	Jupiter
32	96	100	95.39	Saturn
64	192	196	191.91	Uranus*
128	384	388	300.61	Neptune*
256	768	772	395.29	Pluto*

* These planets were discovered after Bode's Law was formulated.

Uranus was the first to be discovered after Bode's Law: William Herschel identified it in 1781 and the orbital radius is quite a good match to the Bode's Law prediction of 196. Around this time, people began to suspect that the Bode number 28 must indicate the presence of an as yet undiscovered body between Mars and Jupiter. On January 1, 1801[1] Giuseppi Piazzi, working in Sicily, discovered a faint starlike object, which was moving in an orbit between Mars and Jupiter. The object was promptly named 'Ceres' and it was the first asteroid (which means 'little star') to be discovered. However, so many more were soon found and they were dubbed the 'vermin of the skies' by some disgruntled watchers.

Although Bode's Law was used by both Adams and Leverrier in their independent calculations to predict the position of Neptune, the distance of that planet does not correspond well with any Bode sequence number.

Using similar, but more complex calculations, Lowell and Pickering predicted a trans-Neptunian planet. It is clear, now, that Pluto with its tiny mass could never have influenced Neptune's path in the manner upon which the calculations were based. So, again, the astronomers found the object, but it wasn't the one they were after.

Clyde Tombaugh discovered Pluto on February 18, 1930 in the constellation of Gemini and it was in that same constellation that William Herschel had found Uranus on March 13, 1781. That is far from the end of the interesting oddities in this story. Pluto had just moved out of the dense starfields of the Milky Way, where it would have been harder to pick it out. Tombaugh, a most vigilant and thorough observer, might have found it even then, but many would not, I am quite sure. Pluto was also close to the ecliptic for the first time in over 200 years. Tombaugh's search was somewhat delayed by the necessity for the completion and installation of a new instrument at the Lowell Observatory near Flagstaff, Arizona – a fortunate delay, for once.

Pluto bears much the same relationship to the bodies in the Kuiper Belt as Ceres does to the asteroids of the Main Belt. Each is the largest by a good margin and each probably contains around 30 per cent of the total mass. Putting Pluto in that category, rather than a genuine planet, allows me to pick out another coincidence. There are four terrestrial planets with an outer belt of smaller bodies of fairly similar composition. Then there are four giant planets with an outer belt of bodies, which appear somewhat similar to the moons of the giants, if not to the giants themselves.

Years ago, I got interested in how astronomers built up a model of our Galaxy. This is quite a task and I am fond of comparing it to trying to map London without moving from my own house in Tottenham. I wonder whether we would have got so far with our understanding of our Galaxy, if the nearby galaxies had been elliptical or irregular. M31, M33, M51 and M81 are all spiral galaxies of a variety of types (plus all four are odd numbers and three out of four are something-ones – logging coincidences gets pernicious after a while). The good luck in being in a part of space where the major galaxies are all of the same broad type as our own is worth a few minutes' thought. There is no doubt that having a good comparison helps, when you are trying to make a model of what you see.

M51 was the first of what were then called 'spiral nebulae' to be identified and that was done by Lord Rosse with his 72-inch reflector in the grounds of Birr Castle in Ireland. This telescope has now been renovated and can be seen by the public.

Thinking of our Galaxy as the 'Milky Way' – a beautiful sight now difficult to see in our light-polluted skies – I found it rather pleasing that the centre of the Galaxy is in Sagittarius and the nearest part of the rim in Gemini. Both of these are zodiacal constellations, but the rest of the Milky Way is aligned at quite a steep angle to the zodiac and goes through such constellations as Cassiopeia.

We hear a great deal about ancient astronomers. I found it interesting to run my computerized star-charts back in time and see what the skies of ancient Egypt and Sumer would have been like. I discovered that at certain times of year those latitudes are ideal for seeing the Milky Way as a band right around the horizon[2], which must be quite a sight!

The dates of Galileo and Newton are interesting: Galileo Galilei 1564–1642; Isaac Newton 1642–1727. It's a bit like the handing on of the Olympic Torch.

It has always struck me as fortunate that Tycho Brahe and Johannes Kepler were able to meet and work together before the fatal dinner in 1601. Kepler arrived in Prague to be Tycho's assistant in 1600. It may also have been fortunate that they did not have to put up with one another for much longer as they did not get on. I doubt if it is a coincidence that Tycho gave the orbit of Mars to Kepler as his first major problem. It was a tremendous task and took Kepler about ten years to accomplish. The outcome was ultimately Kepler's three laws of planetary motion so, coincidence or not, it was fortunate. The orbit of Mars was probably the only one, given the standards of data available, that would have forced Kepler to adopt elliptical orbits.

Before leaving this hotchpotch of oddly related things, I want to mention that a month is roughly 5×6 days and our year is very nearly $3 \times 4 \times 5 \times 6$ days. The great Paul Dirac wrote papers on the coincidences of certain large numbers, which are every bit as interesting, and probably more significant, but I feel a review of his work would require a whole article to itself.

Depending on how you evaluate such propositions, *the* coincidence may be that we are here at all to observe, discuss and enjoy these things. Finally, one last coincidence is my absolute favourite: I was ten years old and had read my first astronomy books (by Patrick Moore) when Sputnik 1 went into orbit. I was just the right age to be interested and inspired.

NOTES

1. In the nineteenth century, they understood dates better, and they knew that this was the first day of their century.
2. Only at certain times of the year, of course.

The Enduring Legacy of Bernhard Schmidt

FRED WATSON

It's probably apocryphal, but it's a charming story – one of the many that circulate in a place like Australia's Siding Spring Observatory. Two German astronomers were visiting for the first time, and asked for directions. 'Please,' they said in perfect English, 'Can you tell us how to find the Smith Telescope?' We can't help but smile at consummate linguistic ability being carried just one step too far.

The 'Smith' telescope they were looking for was the United Kingdom Schmidt, one of the largest examples of this peculiar hybrid class of telescope that all bear their inventor's name. The textbooks call them 'catadioptric' telescopes – the word's Greek roots suggest both the reflection and transmission of light – but no one in the real world calls them that. True, the design incorporates both a mirror to focus the incoming light and a thin glass plate that subtly modifies it before it hits the mirror. But to astronomers, they are just 'Schmidts', and the term embodies all the remarkable properties that make these telescopes special.

PRELUDE TO THE SCHMIDT TELESCOPE

To appreciate those properties and to understand why the Schmidt design was the darling of the astronomical world for more than half a century, you have to cast your mind back a hundred years. In 1902, the remarkable career of the American telescope builder, George W. Ritchey (1864–1945), was just beginning to flourish. Ritchey, then at Yerkes Observatory, near Chicago, had completed a 24-inch (0.6-m) reflecting telescope the previous year, and this modest instrument was providing real competition for the Observatory's giant 40-inch (1-m) refractor – which remains the world's largest refracting telescope today.

Ritchey was interested in the new technique of astronomical photography and, in particular, in photography of nebulae – fuzzy patches of light that were still only poorly understood. He appreciated one of the curious facts of photographic image detection, namely that the speed with which an extended object like a nebula can be recorded depends not on the aperture of the telescope (the diameter of its mirror or lens), but only on its focal ratio, or f/number. This quantity, well known to photographers, is just the focal length of the lens or mirror divided by its diameter.

It seems almost impossible to believe that sensitivity could be independent of aperture, but the effect arises because of the geometry of the situation. When one calculates the amount of light falling on the image of an extended object, the aperture cancels out. The lower the focal ratio, the greater the speed with which the image will be detected – which is why low-focal-ratio optics are often called 'fast'. (The trade-off comes in the scale of the image, which determines the amount of detail it reveals. For a given focal ratio, a smaller aperture results in a smaller image-scale, showing less detail.) Conversely, for point sources like stars, focal ratio is relatively unimportant. Here, aperture is everything.

Ritchey's new 24-inch telescope had a mirror with a focal ratio of f/3.9, astonishingly fast by the standards of the day. By contrast, the 40-inch worked at f/18.6, which was slow even compared with other late-19th-century refractors. Little wonder the new reflector could beat it hands down for nebular photography.

The early decades of the 20th century saw Ritchey's work on large reflecting telescopes come to fruition – his 24-inch was followed by a 60-inch (1908) at Mount Wilson, and then the famous 100-inch (1917) on the same site. The emphasis on nebular astronomy – and in particular the controversy over the true nature of spiral nebulae – demanded power in photographic detection, and spelled the end of the large visual refractor. But these large reflecting telescopes also had an Achilles' heel, and it lay in the paraboloidal shape of their mirrors.

Most people interested in telescopes know that a mirror with a cross-section in the shape of a parabola will form a perfect point image when parallel light (e.g. from a star) falls upon it. But that is only true when the incoming light is exactly aligned with the axis of the parabola. Any departure from this condition causes steadily increasing degradation of the images due to aberrations – mostly coma, which gives stars the appearance of half-open umbrellas or comets (hence the name) with their tails pointing away from the centre of the field of view. With a paraboloidal mirror, coma is noticeable at off-axis angles as small as a few minutes of arc, and becomes progressively more acute with faster focal ratios. Thus, for all the potency of fast reflecting telescopes in nebular photography, they were almost useless for photography over wide angles of view – which is exactly what was needed to record large, faint nebulae.

And there the problem remained. The large reflectors were limited to a narrow field of view. Ritchey himself arrived at a new and elegant solution for wide-field telescopes based on the well-known Cassegrain design (which uses a convex secondary mirror) in the late 1920s, and this extended the achievable field to well over a degree. But for really large-scale photography at fast focal ratios, there was no known optical recipe.

UNUSUAL MAN – UNUSUAL INVENTION

It was an eccentric, ill-adjusted genius named Bernhard Voldemar Schmidt who provided the solution to the problem. This unusual man's short life is one of the more colourful chapters in the history of astronomy. He was born on a tiny, isolated island in the Gulf of Finland on March 30, 1879 into a world that knew little but the day-to-day rituals of farming life and the Lutheran Church. By the time of his death in Hamburg on December 1, 1935, he was celebrated as a master optician and brilliant innovator. A pacifist, he was deeply troubled by events in his adopted country, foreseeing that they would inevitably lead to conflict and the military use of the invention he had cherished for its benefits to humanity. He was right: the Second World War saw both sides using Schmidt-type optics for military purposes.

We know little of Schmidt's early life, but perhaps the best-known episode is the accident in which he lost his right hand and forearm. Somehow, one particular Sunday found him not in church, but in a field experimenting with home-made gunpowder. We are told that the resulting explosion distressed him less for his horrific injury than for the damage it had done to his Sunday suit – and the potential reaction of his parents. Whatever the truth of the story, Bernhard seems not to have let his disability stand in his way during later life.

Schmidt studied optics at the Institute of Technology in Gothenburg, Sweden, and then, at the turn of the century, moved to the Jena district of Germany. This was the epicentre of the optical world; it was not only the home of the Carl Zeiss factory, but also of the University of Jena where Ernst Abbe – one of the greatest physical scientists of the 19th century – was Professor of Optics (see 'The Dawn of Binocular Astronomy' in the *2001 Yearbook of Astronomy*). Schmidt shunned the regimented formality of the academic environment, however, preferring to support himself by manufacturing small numbers of high-quality parabolic mirrors for both amateur and professional astronomers.

It was his undoubted talent as a practical optician that eventually brought Schmidt to the attention of the Hamburg Observatory – and to its director, Dr Richard Schorr. Perceiving the benefits of having such an able craftsman on his staff, Schorr persuaded Schmidt to come to Bergedorf (home of the Hamburg Observatory) and work for him as a 'voluntary colleague'. It was a curious arrangement, for it allowed the eccentric and unruly Schmidt to come and go as he pleased, but still gave the astronomers at Hamburg access to his talents – when he felt like working. As Paul Hodges, a later commentator on Schmidt's work, noted, 'it might have cost Schorr his job and his reputation, but instead it gave the world the Schmidt telescope.'

It was during this time that Schmidt first interacted with Walter Baade (1893–1960), one of the greatest astronomers of the mid-20th century. Baade was then at Hamburg, and he urged Schmidt to explore the possibilities for designing a fast reflecting telescope with a truly wide field of view. When Schmidt eventually succeeded in 1930 Baade was delighted, and it is due to him that the design received the recognition it deserved. Not only did Baade ensure that Schmidt's name was forever linked with the new instrument, but when he later moved to the USA, he championed the design among American astronomers. They, in turn, adopted it with great enthusiasm.

What was this wonderful invention that so neatly solved the problem of recording faint nebulae over many square degrees of sky? In fact, it was simplicity itself, and the reasoning behind it showed Schmidt's crystal-clear understanding of the optical issues at stake. Noting that off-axis coma limits the usefulness of a paraboloidal mirror, he threw away the requirement for it to be parabolic and looked instead at spherical mirrors – ones whose cross-section is simply an arc of a circle. Such mirrors cannot be used in conventional reflecting telescopes because they suffer from spherical aberration – a defect that prevents them from forming sharp images of distant objects.

Schmidt realized that if the incoming beam were limited by an aperture, or 'stop', placed at a spherical mirror's centre of curvature, there would be no preferred axis and therefore no coma. Only spherical aberration would remain, but – crucially – that remains more-or-less the same at all off-axis angles. Therefore, if you could correct the spherical aberration, you would have your perfect wide-field system.

And correct it he did. Schmidt's breakthrough was to replace the stop with a thin glass correcting plate that would introduce into the incoming beam just enough spherical aberration to balance exactly that of the mirror. It was a brilliant solution, and proved to be highly successful in practice. Schmidt spent time seeking an elegant way of producing the shallow, toroidal surface needed on the correcting plate. The method he eventually arrived at involved bending the flat glass plate under vacuum before grinding and polishing it, but that technique was not used to any great extent in subsequent years.

Schmidt built a prototype with an aperture – the diameter of the correcting plate – of 14 inches (0.36 m). It had a focal ratio of f/1.7 and a field of view 12 degrees across – a combination of parameters that was nothing short of stunning. And it worked superbly. At a stroke, Bernhard Schmidt had provided the world of astronomy with exactly the tool it needed to explore the sky at faint light-levels over wide areas, a tool that eventually revealed new and unsuspected variety in the cosmic zoo. Sadly, the prototype did not survive for posterity. Like the peaceful aspirations of its creator, it was destroyed in the Second World War.

It was Baade, once again, who joined forces with Schorr to urge the shy and retiring inventor to publish his design. The result was Schmidt's one and only scientific paper, published in the Communications of the Hamburg Observatory in 1932. Its unassuming title, 'Ein lichtstarkes Komafreies Spiegelsystem' ('A Rapid Coma-free Mirror System') betrays little of the significance of the invention described within it. And in its brief compass, the paper not only sets out the design, but explores several related technical issues – some of them of surprising subtlety.

SCHMIDTS RULE THE WORLD

Thanks to Baade's enthusiasm, Schmidt telescopes proliferated rapidly in the USA during the 1930s. An account published at the beginning of 1941 lists more than 40 existing or projected instruments in the USA and Canada. A handful of them were quite substantial projects. Fewer were built in Europe, perhaps as a result of old-world conservatism.

One of the instruments in the 1941 list is the 48-inch (1.2-m) Palomar Schmidt, which was eventually completed in 1948. This telescope did more than perhaps any other to promote and publicize the usefulness of Schmidt's design. It was built at Mount Palomar Observatory alongside the 200-inch (5.1-m) Hale telescope, for two and a half decades the world's largest optical telescope. The symbiosis between these two instruments is legendary, the 'scouting' capabilities of the Schmidt allowing astronomers to forage for exciting new objects which could then be followed up in more detail with the 200-inch. As we shall see, the Palomar Schmidt also engaged in photographic surveys of the whole northern sky, allowing astronomers free access to sky atlases of unprecedented detail and depth.

Table 1. The World's Largest Schmidt Telescopes

Name	Location	Aperture (m)	Completed
LAMOST	China	4.00*	2004 (estimated)
Tautenburg	Germany	1.34	1960
United Kingdom	Australia	1.24†	1973
Oschin (Palomar)	USA	1.24†	1948
Kiso	Japan	1.05	1976
Byurakan	Armenia	1.00	1961
Kvistaberg	Sweden	1.00	1963
ESO	Chile	1.00†	1972
Llano del Hato	Venezuela	1.00	1978

* Mean effective aperture of movable (reflective) corrector

† Achromatic corrector

In the wake of the Palomar Schmidt (which is now known as the Oschin Schmidt), other observatories invested resources in building large Schmidt telescopes to expedite their own surveys. The nine largest – with corrector plates of 1 m or more in diameter – are listed in Table 1. Apart from the proposed LAMOST telescope (of which more later), these telescopes all date from a 30-year period between 1948 and 1978. Those were perhaps the golden years of the photographic Schmidt telescope and, truly, all these instruments are products of their time.

In their construction, these so-called 'classical' Schmidts all follow a similar generic pattern (see Figure 1). The thin glass plate – the 'Schmidt corrector' – is the heart of the system. It is usually flat, but is

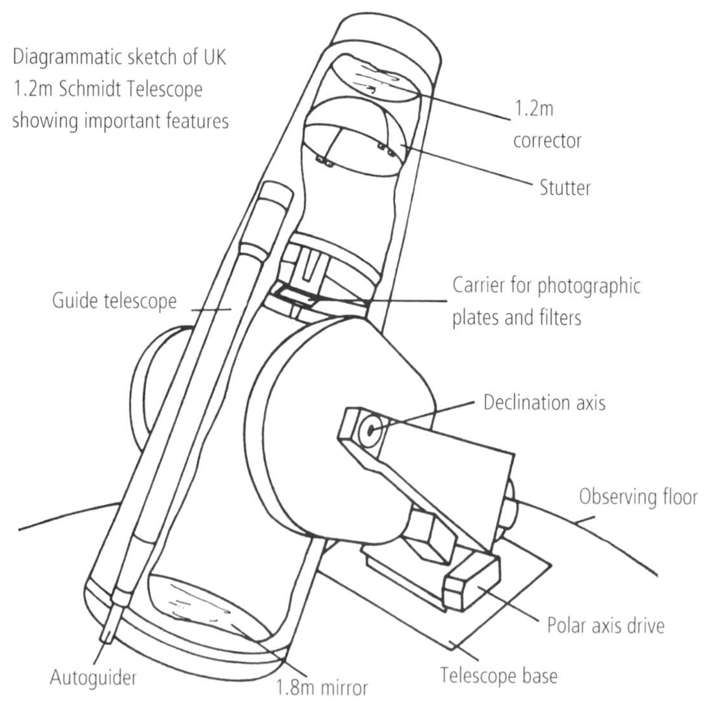

Figure 1. Construction of a large Schmidt telescope. The light passes through the corrector at the centre of curvature of the mirror, which forms an image halfway up the tube. Today, the photographic plate has been largely replaced by TV-type CCD cameras or optical fibres that feed light out of the telescope to a spectrograph. [Courtesy Royal Observatory Edinburgh]

sometimes bent to a meniscus shape to prevent the formation of misleading ghost images. The complex, shallow profile superimposed on it (sometimes on both its surfaces) has to be precisely computed to match the spherical aberration of the mirror. Because the plate is not a lens as such (for it is the mirror that does the focusing), flexure in the glass presents little difficulty, and Schmidt correctors can be made bigger than the lenses of refractors. For the most exacting work with large Schmidt telescopes, the dispersion of light in the material of the corrector (its separation into rainbow colours) becomes a problem, and an achromatic (two-component) corrector becomes necessary.

The image in a Schmidt telescope is formed on a curved surface which is a segment of a sphere concentric with the mirror. To match this surface, the photographic film or plate (made of very thin glass) must be deformed by clamping it in a suitably shaped plateholder. Amazingly, the glass plates seldom break. The curved focal surface is one of the few real drawbacks of the Schmidt system; another (which is important from the point of view of cost in large instruments) is that the telescope tube must be made twice as long as its focal length to support the corrector at the mirror's centre of curvature. In general, this means that enclosures for Schmidt telescopes have to be twice as large as their counterparts for equivalent conventional telescopes.

One other subtlety of the Schmidt design – and one that was fully appreciated by Schmidt himself – is that the mirror must be made larger than the corrector if the field of view is to be uniformly illuminated. To understand why this is the case, imagine them both the same size. A cylinder of light passing obliquely through the corrector (as it does for off-axis objects) will fall on only part of the mirror surface, so the corresponding part of the image will receive less light. In the Palomar and United Kingdom Schmidts (which have almost identical optical specifications), the mirror is 72 inches (1.8 m) in diameter and will accept the full 48-inch (1.2-m) diameter beam delivered by the corrector up to an off-axis angle of almost 3 degrees. Beyond that, there is a slight reduction in the illumination of the image, although the image-quality remains superb. Both these telescopes take photographs 6.6 degrees on a side on 14-inch (0.36 m) square plates or films. That is big enough to allow the UK Schmidt to capture the entire Southern Cross in a single exposure.

Schmidts have also proliferated in smaller sizes and, perhaps even more notably, in hybrid forms. The Schmidt-Cassegrain design,

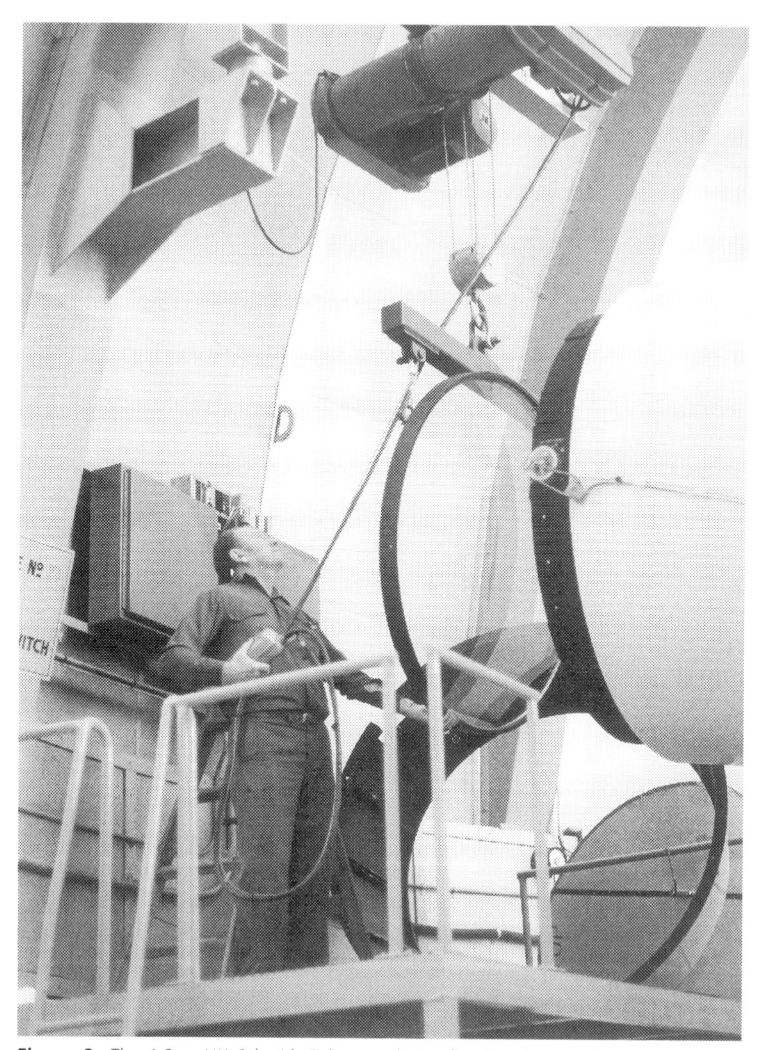

Figure 2. The 1.2-m UK Schmidt Telescope being fitted with one of its two objective prisms. Used singly or in combination, these narrow-angled glass prisms turned each image on the wide-angle photograph into a miniature spectrum. They were useful for finding rare objects like quasars or Wolf-Rayet stars. [© Royal Observatory Edinburgh]

for example, which combines advantages from both systems, was pioneered in large sizes by the British optician Robert Waland at the University of St Andrews. (Your humble author used Waland's prototype 15-inch telescope for his MSc project back in the psychedelic 1960s.) And, in recent years, the small Schmidt-Cassegrain has become the 'standard' class of telescope for serious amateur astronomers. No doubt Bernhard himself would be delighted to see his principle being used to bring the joys of hands-on astronomy to so many people.

Before we move on to the kinds of astronomy carried out with large Schmidt telescopes, one other facet of the story might be mentioned: the Schmidts that never were. The 1941 list mentions a 60-inch (1.5-m) instrument for the Harvard Observatory that was said to be in construction at the time by the Fecker Company; it seems to have disappeared without trace. And did you know that the 2.5-m Isaac Newton Telescope in La Palma was, in a previous incarnation, intended to be a Schmidt telescope? The original 98-inch Pyrex mirror for this telescope was polished spherical by the optical company of Grubb Parsons in the 1950s for a hybrid convertible Schmidt to be located at the Royal Greenwich Observatory at Herstmonceux in Sussex. Priorities changed and the plan was abandoned, the INT eventually gracing the Sussex countryside as a Cassegrain instrument. But for many years the unfigured Schmidt corrector plate gave sterling service as a coffee-table top in the RGO's spacious foyer.

CLASSICAL SCHMIDTERY . . .

It was probably Victor Clube, a prominent astronomer then at the Royal Observatory in Edinburgh, who coined the term 'Schmidtery' during the 1970s. Typical of this scientist's entertaining wit, it was an apt term to describe the unique kind of astronomy that can be carried out with a photographic Schmidt telescope. In those days, it was very much in the ascendant at Edinburgh, which was the headquarters for the UK Schmidt Telescope (UKST) in Australia.

More than anything else, it was the Schmidt telescope's suitability for observing very large areas of sky in great detail and with high sensitivity that captured the imagination of astronomers. Not only were large extended objects like nebulae and comets perfect targets, but the telescope's ability to record large numbers of more compact sources

Figure 3. Blue-light photograph of the Pleiades star-cluster taken with the UK Schmidt Telescope. The marvellous blue nebulosity that characterizes the cluster is clearly visible. Each bright star image is accompanied by artefacts: the diffraction spikes come from the cross-structure supporting the plateholder, while the sharply defined halo is a ghost image caused by light being reflected between the two surfaces of the corrector. Note also the left-to-right satellite trail. [© Royal Observatory Edinburgh/Anglo-Australian Observatory]

such as stars and galaxies had great potential for statistical studies. Even a single photograph taken with the UK Schmidt records typically half a million such objects.

On top of this was the Schmidt telescope's capability as a survey instrument to identify unusual or specific kinds of object for subsequent follow-up with larger, conventional telescopes. The relationship between the Palomar Schmidt and the Hale Telescope has already been mentioned, and it was no accident that the UK Schmidt and the 3.9-m Anglo-Australian Telescope (AAT) were built together at Siding Spring Observatory in the early 1970s. Unlike the Palomar telescopes, however, these two instruments were under separate management until the UKST became part of the Anglo-Australian Observatory (AAO) in 1988.

During the second half of the 20th century, these various aspects of Schmidtery were rolled together in several major surveys of the sky undertaken with three of the world's largest Schmidts – the Palomar, ESO and UK Schmidts. They resulted in complete atlases of the sky at a variety of visible-light wavelengths and at several different epochs of time. Blue, green, orange, red and near-infra-red sky surveys were carried out. Probably the most famous of them were the original surveys – the *National Geographic*/Palomar Observatory Sky Survey in the north (1950s), and the European Southern Observatory/ Science and Engineering Research Council Southern Sky Survey (1970s/80s). These have been repeated in recent years to provide time-resolved data.

Further, limited-area photographic surveys have also been carried out at other wavelengths. One of the most ambitious of these – and one of the last to be completed – was a survey of the southern Milky Way and the two Magellanic Clouds (our Galaxy's biggest satellites) in the red light of hydrogen at a wavelength of 656 nm with the UKST. It will just be finished as this edition of the *Yearbook of Astronomy* goes to press.

The discoveries from these surveys embraced most areas of astronomy – from Solar System studies (comets and asteroids) to the structure of our own Galaxy, and from superclusters of galaxies to the large-scale structure of the Universe. Some of them have been pivotal in our present understanding of the cosmos. But to say that Schmidt photography alone has produced these advances would be an over-simplification, because there is one other ingredient of technology that has allowed the photographs to be fully exploited.

If you were to be confronted with a photograph from the UKST and told to do star-counts on it, you would find it a daunting prospect. No one would relish the task of sorting through hundreds of thousands of images, no matter how refined the microscope you had been provided with. It was because of this that high-speed automatic plate-measuring machines were developed, both at Edinburgh and Cambridge, to turn each individual Schmidt photograph into around 2 Gbytes of machine-readable data (see 'The Astronomer's Microscope' in the *1999 Yearbook of Astronomy*). The current machine at Edinburgh – SuperCOSMOS – can accomplish that task in little more than a couple of hours.

The way is then open for the formidable computing resources of the modern astronomer to be brought to bear on photographic Schmidt material. Sifting objects by colour (for quasars or brown dwarfs, for example), by movement (for asteroids or high-proper-motion stars, for example) or by variability then becomes trivial and gives astronomers almost unconfined powers of detection within the limits of the original photographic material. Thus have some of the most spectacular recent discoveries been made from Schmidt telescopes. Even simple star-counts can produce exciting results, as in the subtle enhancements of star numbers – invisible to the eye – that have recently betrayed the presence of dwarf galaxies behind the Milky Way.

A few tasks still need human eyes, however. Eyeball scans of the new UKST hydrogen-light survey, being carried out by Edinburgh astronomer Quentin Parker with Malcolm Hartley of the AAO, are revealing hitherto-unknown southern planetary nebulae in great numbers. And, almost before each new film from the telescope is dry, near-Earth-object specialist Rob McNaught scans it for fast-moving asteroids. He has no time to wait for the films to be sent to Edinburgh to be measured – by then the possibility of recovering any new objects would be long gone.

. . . AND ITS DECLINE

Digitized Schmidt sky-survey data provided by SuperCOSMOS and other measuring machines have been available on-line on the World-Wide Web and on CD-ROMs for some time now. These data-sets give instant access to the sky, and nowhere are they more useful than in the control-rooms of the world's 4-m class telescopes, where they allow astronomers to check quickly and easily the identification of their target objects. The brightness limits of the sky surveys – around magnitude 22 in the blue waveband, for example – are well matched to the limits of the 4-m telescopes when used in spectroscopic mode, so most potential targets will be found in the surveys.

But things are changing. Today, we no longer live in a world where the largest optical (visible-light) telescopes are 4-m class instruments like the AAT. The new workhorses of astronomy are in the 8- to 10-m class – the Geminis and Kecks of this world (see 'Optical Astronomy, the Early Universe and the Telescope Super-league' in the *2000 Yearbook of Astronomy*). The capabilities of these instruments take them beyond the limits of the existing sky surveys in brightness. And even bigger telescopes – the so-called Extremely Large Telescopes of the

25-m class – are undoubtedly on the way. So what are astronomers doing about finding their target objects?

The answer to this question is not simple. Prospects for surveying the entire sky to fainter limits than the existing sky surveys are poor, a conclusion dictated not just by technological considerations but also by more fundamental limitations. In principle, it would be possible to equip the world's large Schmidts with mosaics of CCDs – charge coupled devices, whose sensitivity in recording images is some 20 times greater than photography. While that would certainly take the sky surveys deeper, it would not yield the accompanying improvements in resolution (spatial detail) that the big new telescopes demand. Most Schmidts are limited by their optics and local atmospheric turbulence to a resolution of an arc second (1/3600 degree) or more, and that is simply not fine enough for 8-m class telescopes. Moreover, much of the observing being carried out with the new telescopes ventures too far into the infra-red for any possible contribution by Schmidt telescopes to be useful.

These considerations have recently led British astronomers to invest almost £25 million in a new 4-m class wide-field telescope called VISTA. It is not a Schmidt; its design gives it a field of view of only 1.5 degrees in the visible waveband and one degree in the infra-red. But its resolution of around 0.25 arc seconds is much better matched to the 8-m class telescopes. While VISTA will carry out large-scale surveys of particular areas of sky, it is unlikely ever to take on the task of mapping the whole sky. That demands Schmidt-sized fields of view. So VISTA and other proposed large survey telescopes can only partially satisfy the requirements of the brave new world.

Meanwhile, the operators of large Schmidts have been looking very hard at the future roles of their telescopes. Their deliberations began in earnest nearly a decade ago in March 1994, when almost a hundred of the world's astronomers attended an international conference called 'The Future Utilization of Schmidt Telescopes' in Bandung, Indonesia. It was a very lively meeting, suitably spiced with Indonesian cuisine and hospitality, and it had about it an air of optimism that promised a rosy future for the Schmidt. While that might have eventuated in the short term, it seems a naïve view from today's perspective.

Since then, the impact of the new large telescopes has been felt throughout astronomy, and several Schmidts that have come to the end of their photographic careers have simply been mothballed. Most notable among them is the 1-m ESO Schmidt. A few others have been equipped with experimental CCD cameras, a strategy pioneered with 0.6-m and 0.9-m class instruments in China, France and the USA. Astronomers at the Kiso Schmidt in Japan not only deployed a CCD camera but also a near-infra-red imager, and were limited in their success more by the poor observing conditions at Kiso than by the technology. Happily, the Oschin Schmidt has remained operational, and has been turned over to the search for near-Earth objects using CCD imaging. For this kind of work, Schmidts are second to none, and it is clearly a worthwhile use of the telescope. It would be the ultimate proof of the efficacy of Bernhard Schmidt's design if, one day, it saved the world.

SCHMIDTERY RE-INVENTED

For all it might have rescued some of them from closure, the use of CCD cameras on large Schmidt telescopes represents an under-utilization of their potential. CCDs are still unable to compete with photographic film in area of coverage, and Schmidt telescopes are essentially large-format devices. The CCDs might only access a degree or two of a telescopic field-of-view perhaps ten degrees across. And so the one attribute that makes Schmidt telescopes unique – their ability to engage in all-sky survey observations – is lost.

What sort of alternative strategy would enable astronomers to regain that potential and transfer it from the photographic era into today's information age? Back in the late 1960s, a British information theorist called Peter Fellgett had the right idea. He suggested that valuable multi-object photometry could be carried out by equipping a large Schmidt telescope with scores of light-sensing avalanche photo-diodes distributed over the focal surface to match the arrangement of target stars. That suggestion never went anywhere, but it was a precursor of one that did.

A dozen or so years later, in 1982, two astronomers at the UKST proposed that optical fibres could be used to collect the light from individual targets in the field of view of the Schmidt and transmit them to a spectrograph for analysis. One of them was John Dawe, the other was a fellow called Fred Watson. The basic idea was not new; three years earlier, a group in the US had demonstrated the multi-fibre technique with the Steward Observatory's 2.3-m telescope at Kitt Peak. And Peter Gray, a capable and enthusiastic engineer at the AAO, had applied it to the AAT in 1981 (see 'Astronomy's multi-fibre revolution' in the *1995 Yearbook of Astronomy*). Dawe and Watson, however, were the first to point out the significance of field-of-view in using the technique – and its consequent potential for Schmidt telescopes. The ability to collect spectra of many objects at a time spread over a wide field promised to be immensely powerful.

Twenty years on, the UKST has become a telescope that spends virtually all its time operating in this mode. The technique has developed over the years from an experimental sideline in the 1980s through to its use for about 30 per cent of observing time during the1990s, and is now set to take the telescope to 2006 and beyond. New technology has just been introduced that automates the fibre positioning process and allows the telescope to observe 150 objects at a time. The prospect of gathering the spectra of 1000 galaxies on a single winter's night is now quite real.

What has spurred the investment in this technology at the UKST – resulting in a $A900,000 instrument called 6dF (for six-degree field, after its big brother 2dF on the AAT) – is, once again, the prospect of an all-sky survey. This time, though, it is spectroscopic, and its aim is to measure the redshifts (and hence the three-dimensional positions) of some 120,000 galaxies spread over the whole southern sky. That will give a clear picture of the distribution of galaxies in the local Universe over an entire hemisphere – a half-apple-shaped volume of space extending out almost a billion light-years.

The 6dF Galaxy Survey has another ambition, though, and if it succeeds it will be a real coup for a 1.2-m telescope – even one as special as the UKST. By using alternative methods of obtaining distances for a local subset of these galaxies (about 15,000 of them), it will be possible to estimate their actual space velocities over and above their motion due to the expansion of the Universe. That will allow the determination of large-scale mass flows, and give a direct measurement of the distribution of matter – both visible and invisible – in our bit of the Universe. It is something that can only be done by looking at a whole hemisphere of the sky. And therefore, of course, it is something that can only be done with a Schmidt.

One or two other Schmidt telescopes have followed in the footsteps of the UKST and implemented multi-fibre systems. Most notable is the world's largest Schmidt, the 1.3-m telescope at Tautenburg not far from Ernst Abbe's Jena. Unfortunately, this instrument is on a very poor site, and its 2.3-degree-field 34-fibre TAUMOK system has never realized its full potential.

Of far greater significance is a quite unexpected development in the world of Schmidtery. This is a new Schmidt – dedicated entirely to multi-object spectroscopy – being built on a scale that makes existing Schmidt telescopes look like toys. Its name is LAMOST, an acronym for Large Area Multi-Object Spectroscopic Telescope, and it is being built by the Chinese Academy of Sciences for the Xinglong Station of Beijing Astronomical Observatory. Completion is expected sometime after 2004.

Optically, this telescope is a classical Schmidt, but by any standards it is a most un-Schmidt-like instrument (see Figure 4). Its main mirror is spherical, as usual, but it consists not of a single piece of glass, but of 37 hexagonal segments about 1-m across, arranged in the style of the mirrors of the two Keck Telescopes. The resulting primary mirror is 6.67 m × 6.05 m in size. This mirror is fixed, and points downwards at

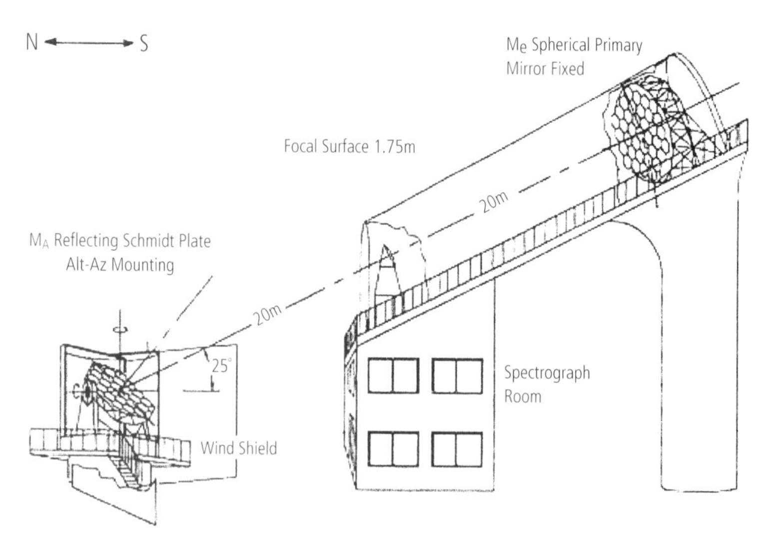

Figure 4. Schematic diagram of LAMOST, the all-reflecting 4-m Schmidt telescope being built in China. The reflective corrector plate also serves as a steerable mirror to direct light onto the 6-m primary. [Courtesy LAMOST Project Management Team]

an angle of 25°. What it looks at is not the ground, but a second, almost flat mirror that can be steered around to direct light from anywhere in the sky onto the spherical mirror.

The really cunning part is that this flat mirror is actually the Schmidt corrector, for it carries the shallow profile that corrects the spherical aberration of the primary mirror. It, too, is segmented, this time into 24 hexagons that combine to make an area of 5.72 m × 4.40 m. The effective aperture of the telescope depends on the angle between the flat mirror and the inclined axis, but it averages out at about 4 m. There is one more subtlety associated with the flat mirror – one that brings a real challenge to the technology. As the mirror moves to target new field-centres or to track them across the sky, the Schmidt profile must change to maintain the correction of the primary mirror. So each segment of the flat will be deformed by actuators that bend the glass slightly to achieve the desired shape. This kind of engineering – while already proven in existing active-optics systems – is far from straightforward.

Halfway between the flat and the primary mirror is the focal surface and, like everything else about LAMOST, it is on a massive scale. Its 1.75-m diameter covers a field of view of 5°, and the Chinese expect to populate it with no less than 4000 fibres – each of which will carry the light of a single target to a multi-object spectrograph. The automatic positioner for these fibres will itself be a major engineering feat, but the results promise to be very spectacular.

LAMOST will combine the penetrating power of 2dF on the AAT – which is surveying a quarter of a million galaxies in a pie-shaped slice of the Universe out to about 2 billion light-years – with the all-sky capability of 6dF. The expected result, for galaxy surveys at least, is a three-dimensional map of the positions of tens of millions of galaxies. And like 6dF and 2dF, LAMOST will not be limited to galaxy surveys. The new telescope is bound to revolutionize our understanding of many fields of astronomy – galactic as well as extragalactic – by making new all-sky surveys. Readers of the *Yearbook of Astronomy* will no doubt watch this space with interest.

THE ENDURING LEGACY

Would Bernhard Schmidt have recognised LAMOST had he lived to see it? You bet he would, for the design speaks eloquently of the effectiveness of his optical system. One can imagine the delight he would have taken in the variable surface that maintains the profile he himself first computed. On a vastly smaller scale, it's tempting to believe that he would have taken delight in the UKST's 6dF, too. For here, not only is the telescope that feeds the fibres a Schmidt, but the main components of the spectrograph (the collimator and camera) are also Schmidts – baby ones, with apertures of 150 mm. The light from each galaxy measured by 6dF traverses no less than three Schmidt systems on its way to the detector.

With the promise of LAMOST, the success of the revitalized UKST, and the ongoing exploitation of the Oschin Schmidt in NEO research, there is still much for the world's great Schmidt telescopes to do. Bernhard Schmidt's legacy has stood the test of time. Its endurance is by no means exhausted yet.

ACKNOWLEDGEMENTS

Rather a lot of my career seems to have been concerned with Schmidt telescopes, and it is a pleasure to take this opportunity to thank all those who've helped along the way. They include present and former staff members of the UK Schmidt Telescope in Australia and its former headquarters in Edinburgh (now the Wide-Field Astronomy Unit of the University of Edinburgh), as well as the scientists concerned with the measuring machines: COSMOS and SuperCOSMOS in Edinburgh, and APM in Cambridge. An immense contribution has also been made by the staff of the Anglo-Austrian Observatory, including David Malin who – while never himself a UK Schmidt observer – has worked wonders in promoting the photographic output of the telescope.

Finally, I am indebted to Patrick Moore for his suggestion that an article about Schmidts would be a 'good thing' for the *2002 Yearbook of Astronomy*.

A Universe of Darkness

IAIN NICOLSON

Within range of large modern telescopes, which can probe out to distances in excess of ten billion light-years, lie billions of galaxies, each of which contains billions of stars. Galaxies are clumped together into groups, which contain from a few to a few dozen members, or clusters, which may contain hundreds, or even a few thousand member galaxies. Clusters themselves are loosely aggregated into huge, straggly, superclusters, with diameters of a hundred million light-years or more.

According to the widely accepted Big Bang model, the Universe originated some 15 billion years ago in a hot explosive event. Space has been expanding, and galaxies receding from each other, ever since. The theory has been very successful in explaining several key features of the observable Universe, in particular, the recession of the galaxies, the relative abundances of the lightest chemical elements, and the fact that space is filled with a faint background of microwave radiation, radiation that was released about 300,000 years after the start of the expansion, when space, for the first time, became transparent.

WILL THE UNIVERSE EXPAND FOREVER?

Intuitively, we would expect that gravity would cause the expansion to slow down as time goes by. In a universe dominated by matter, if the overall mean density of the Universe exceeds a particular value, called the critical density, gravity will eventually win the battle and at some time in the distant future the expansion will cease. Thereafter, slowly at first, then ever more rapidly, galaxies will begin to fall together until everything piles up into a Big Crunch. The best current estimates indicate that the critical density is about 9×10^{-27} kgm^{-3} , equivalent to an average of about five hydrogen atoms per cubic metre of space. If the mean density is less than the critical value, the expansion will slow down, but will continue at a finite rate forever. If the mean density is precisely equal to the critical density, the Universe will just, but only just, be able to expand forever, the speed of recession of the galaxies approaching ever closer towards zero as time goes by.

A Universe that expands forever is called 'open', whereas a Universe that expands to a finite volume then collapses, is called 'closed'. A Universe in which the mean density is exactly equal to the critical density, and which 'sits on the fence' between the open and closed models, is called 'flat'.

The presence of matter and energy causes space to be curved. In a closed Universe, space is positively curved – wrapped round on itself rather like the surface of a sphere. In a positively curved space, rays of light travel along curved paths and parallel lines drawn at any particular location eventually intersect (just as lines of longitude – or meridians – intersect at the poles of the Earth). In an open Universe, space is infinite in extent and is negatively curved – like the surface of a saddle; parallel lines diverge from each other in a space of this kind. In a flat Universe, the net overall curvature is zero; rays of light travel in straight lines (apart from being deflected by localized clumps of matter), parallel lines remain always separated by the same distance, and space, again, is infinite in extent.

According to the inflationary hypothesis, first proposed in the early 1980s by Alan Guth, the Universe underwent a brief but dramatic phase of accelerating expansion – called inflation – very early in its history. During this brief bout of accelerated expansion all distances in the Universe increased by a huge factor (at least 10^{50}). The inflationary hypothesis has been remarkably successful in explaining why the microwave background is smooth and uniform over the whole sky (temperature variations from place to place on the background sky are only about one part in a hundred thousand) and in showing how tiny variations in density could have provided the seeds from which galaxies and clusters were formed. Inflation would have blown up the Universe to such a vast size that its overall curvature would have become indistinguishably close to zero. Because the inflated Universe is so huge compared to the region we can observe (the observable Universe) space appears flat, just as the Earth appears flat if we look only at a tiny part of its surface. Inflation, therefore, would produce a Universe with a geometry that was flat, or indistinguishably close to flat, and with a mean density indistinguishably close to the critical density.

Many theoretical cosmologists are attracted by the idea of a flat-space

Universe with a density equal to the critical density. What, however, do the observations show?

LUMINOUS MATTER IS NOT ENOUGH

In principle, the mean density of luminous matter – the visible stars and gas clouds that make up the galaxies – can be determined adding up the masses of the luminous constituents of galaxies and clusters within a large enough volume of space to provide – hopefully – a representative sample of the Universe as a whole. Measurements of this kind show that luminous matter contributes only about 0.5–1 per cent of the critical density. If luminous matter were the sole constituent of the Universe, it would certainly be open and fated to expand forever.

However, there is strong observational evidence to show that galaxies and clusters, and the Universe as a whole, contain far more dark matter – matter that does not radiate detectable quantities of electromagnetic radiation – than visible matter. The 'dark stuff' may play a crucial role in determining the geometry and evolution of the Universe.

EVIDENCE FOR DARK MATTER IN GALAXIES AND CLUSTERS

Evidence for the existence of dark matter is provided by the rotation curves of galaxies and the internal dynamics of clusters of galaxies, with further support coming from gravitational lensing.

In spiral galaxies such as our own Milky Way system, most of the visible matter is concentrated in a central bulge. If most of the mass were concentrated where most of the light is (in the central bulge), the speeds at which stars and gas clouds beyond the bulge revolve around the galactic centre should decrease with increasing distance in a similar fashion to those of the planets in their orbits around the Sun. The galactic rotation curve (a plot of rotational velocity against distance from the centre) should slope downwards.

In fact, the observed orbital velocities of stars and gas clouds in spiral galaxies remain virtually constant (or even increase with distance) out to and beyond their visible boundaries. Because extra mass is needed to provide the additional gravitational pull that causes distant stars and gas clouds to move as fast as those which are closer to the centre, spiral galaxies have to contain 5–10 times as much dark matter as luminous matter. The usual assumption is that the visible galaxy is embedded within an extensive spheroidal halo of dark matter. In the case of the Milky Way galaxy, observations of the motions of smaller neighbouring galaxies indicate that the radius of the galactic halo is at least 150,000 light-years and perhaps as great as 700,000 light-years.

DARK MATTER IN CLUSTERS OF GALAXIES

Within clusters, the relative velocities of member galaxies are much higher than can be explained by the gravitational influence of luminous matter alone. In order to hold themselves together, and prevent member galaxies from escaping, groups and clusters have to contain 10–50 times as much dark matter as luminous matter.

Further evidence for dark matter in galaxies and clusters is provided by gravitational lensing, a phenomenon which arises because light is deflected by the gravitational influence of large masses. Where a massive object or concentration of matter lies between an observer and a background object, it can act like a lens to produce a distorted image of the background source. Where light from a background cluster passes through a foreground cluster, numerous arc-like images of the background galaxies can be seen. Analysis of the light paths followed by these rays allows the mass of the foreground cluster to be estimated. Typically, gravitational lensing results point to cluster masses that exceed the visible masses by factors of ten or more.

It is also possible that the Universe contains galaxies that are completely dark – which emit no detectable light because they contain few, if any, stars. As possible evidence for this kind of object, Neil Trentham and co-workers, of the Institute of Astronomy, Cambridge, have cited galaxy UGC 10214. A stream of matter appears to be flowing from this galaxy, as if it were interacting with another one, but no such galaxy is visible. The stream of matter appears to be flowing towards nothing.

Figure 1. The galaxy UGC 10214 showing a stream of matter heading towards 'nothing'. [Courtesy of Simon Hodgkin and Neil Trentham, Institute of Astronomy, Cambridge]

THE NATURE OF THE DARK MATTER

Some of the dark matter that resides in galaxies and their haloes may consist of ultra-faint 'stars', bodies with luminosities too low for them to be detected by current techniques. Potential candidates include old white dwarfs that have faded below currently detectable levels, neutron stars, black holes, brown dwarfs and bodies of planetary mass. White dwarfs, neutron stars and black holes represent different end-points for the evolution of stars. Brown dwarfs are cool dim bodies with masses of less than 0.08 solar masses which, because they are insufficiently massive, never attain temperatures sufficiently high to initiate the hydrogen fusion reactions that power normal stars. Bodies with masses below 0.013 solar masses (13 times the mass of the planet Jupiter) are more akin in nature to Gas Giant planets than stars and are regarded, therefore, as 'planets'.

Despite their exceedingly low luminosities, significant numbers of brown dwarfs have been detected in recent years. Furthermore, recent high-sensitivity infra-red studies of the star-forming region in the Orion nebula have also revealed a number of 'free-floating planets', objects with masses in the range 8–13 Jupiter masses which appear to be unattached to any stars. Although these specific results hint that our Galaxy perhaps may contain as many brown dwarfs and 'planets' as stars, because these bodies are so much less massive than stars, their total contribution to the overall mass of the system is likely to be small.

MICROLENSING AND THE SEARCH FOR MACHOS

Dark bodies in the galactic halo, such as subluminous stars, black holes and planets, have come to be known as MACHOs ('MACHO' being an acronym for MAssive Compact Halo Object). During the past few years, several research teams have been searching for MACHOs by looking for the telltale signature of a phenomenon called gravitational microlensing.

If a MACHO passes between the Earth and a distant background star, its gravitational field will act like a lens to produce a magnified image of that star. If the alignment were absolutely perfect, the observer would see the image of the star spread out into a tiny ring of light, called an Einstein ring, the radius of which depends on the mass of the lensing object. In practice, with a lensing object of mass comparable to the Sun,

Figure 2. Hubble Space Telescope image of multiple arc-like patterns within the rich galaxy cluster Abell 2218 caused by gravitational lensing. [Photograph courtesy W. Couch, University of New South Wales, R. Ellis, University of Cambridge and NASA/STScI]

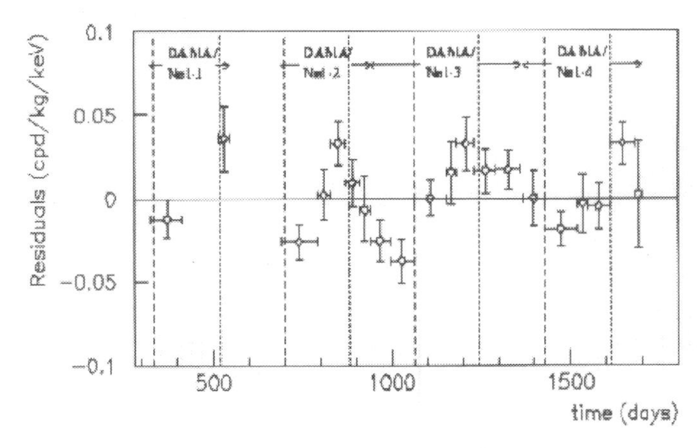

Figure 3. DArk MAtter (DAMA) count rates over a period of four years showing a possible WIMP component. [Courtesy R. Bernabei, Universita di Roma]

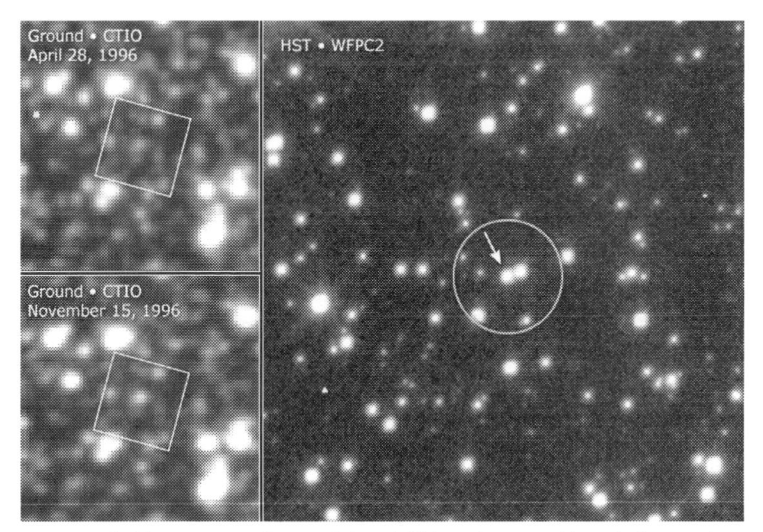

Figure 4. (Left) Two images of a crowded starfield through a ground-based telescope show subtle brightening of a star due to gravitational microlensing. [Courtesy NOAO, Cerro Tololo Inter-American University] (Right) Hubble Space Telescope image of the same field resolves the lensed star and yields its true brightness. [Courtesy Dave Bennett, University of Notre Dame, Indiana and NASA/STScI]

located in the galactic halo, the resulting Einstein ring would be far too small to be resolved by any existing telescope. However, because the gravitational lens concentrates the light of the background star, the star's apparent brightness will rise and fall in a very distinctive way as the lens passes in front of it. The duration of the event depends on the mass, distance and velocity of the lens.

During the past six or seven years, the joint US-Australian MACHO Project has recorded between 13 and 17 well-defined events in the direction of the Large Magellanic Cloud (LMC) – about five times as many as would be expected, statistically, on the basis of the known population of stars along the line of sight between the Earth and the LMC. The durations of these events (34–230 days) correspond to MACHO masses in the region of 0.6 solar masses, masses which match those of white dwarfs rather than those of black holes, neutron stars, brown dwarfs or planets. These results, which are consistent with those obtained by the French EROS project, imply that MACHOs are unlikely to contribute more than 10–20 per cent of the total mass of the dark matter halo.

IS THE DARK MATTER BARYONIC?

Particles such as protons and neutrons, which are acted upon by the strong nuclear force which binds together the nuclei of everyday atoms, are known collectively as baryons. Stars, planets, brown dwarfs, planets, white dwarfs, neutron stars and black holes are composed of baryons, (or in the case of black holes have formed through the gravitational collapse of bodies consisting of baryons) and are, therefore, examples of baryonic matter.

According to standard theory, the present abundances of the lightest chemical elements (hydrogen, deuterium, helium and lithium) were determined by nuclear reactions that took place during the first few minutes of the hot 'fireball' phase of the Big Bang. This process, which is known as 'Big Bang nucleosynthesis' (abbreviation: BBN) depended in a very sensitive way on the mean density of baryonic matter in the Universe. In order for these elements to exist in their presently observed relative proportions, the mean density of baryonic matter can be no more than about 3–5 per cent of the critical density. It is highly unlikely, therefore, that baryonic matter in any or all of its forms can account for all the dark mass that appears to be present in galactic haloes or in galaxy clusters, let alone endow the Universe with its critical density.

NON-BARYONIC DARK MATTER

If the Universe has an average density equal to the critical value, and baryonic matter (luminous and dark) comprises no more than 5 per cent of the total, the remaining 95 per cent of the mass-energy of the Universe has to exist in some completely different form. One possibility is non-baryonic matter – exotic elementary particles that do not respond to the strong nuclear interaction, which hardly ever interact with ordinary matter and which, therefore, are exceedingly difficult to detect. Candidate particles include neutrinos and WIMPs. Neutrinos are known to exist. The existence of WIMPs (an acronym for Weakly Interacting Massive Particles) has been predicted theoretically, but as yet none of these particles has been definitely detected.

NEUTRINOS

Neutrinos are particles with zero electrical charge and zero, or exceedingly tiny rest-masses (if its rest-mass is zero, a stationary neutrino would weigh nothing at all). Theory suggests that a vast population of neutrinos, left over from the Big Bang, permeates the Universe, there being, on average, several hundred million neutrinos in each cubic metre of space. Because they are so abundant, the average neutrino mass need be only about a ten-thousandth of the mass of an electron for there to be enough mass tied up in neutrinos to halt the expansion of the Universe.

Significantly, recent experimental and observational data – in particular, some results obtained by the giant Kamiokande II neutrino detector in Japan – while not providing a direct measure of neutrino masses, do imply that their masses are finite but are probably no more than a millionth of the mass of an electron – far too small for neutrinos to provide more than a tiny fraction of the critical density.

WIMPS

WIMPs are hypothetical particles, with masses ranging from a few to thousands of times the mass of a proton, that interact only very rarely with ordinary matter. The existence of particles of this nature is predicted by certain theories that attempt to link together and 'unify' the fundamental forces of nature. In particular, a theory known as supersymmetry ('SUSY') predicts that at very high energies (such as would have prevailed early in the Big Bang) each ordinary particle has a heavy partner particle. For example, the supersymmetric partner to the photon is the photino, the quark has an associated 'squark', the electron the 'selectron', and so on. The current front-runner among candidate WIMPs is the neutralino, the lightest of the supersymmetric particles, with a probable mass in the range 10–1,000 times that of the proton. Although the more massive supersymmetric particles would have decayed as the Universe expanded and cooled, the neutralino is expected to be stable and, if the theory is correct, to exist in large numbers in the present-day Universe.

Because a WIMP is a very massive particle, when it strikes an atomic nucleus (a very rare event), it causes it to recoil. In principle, the recoil energy can be detected by measuring the heat produced by the event, by detecting the electrical charge liberated when the recoiling nucleus ionizes neighbouring atoms, or by measuring the tiny flash of light that is released when the impacted nucleus recoils. The required sensitivities are exceedingly high and any such experiments are bedevilled by spurious events caused by impacts of cosmic rays (highly energetic charged particles arriving from space) and by natural radioactivity in the laboratory and its surroundings. Although the situation can be improved in various ways, for example, by placing detectors deep underground so that the overlying rock shields the apparatus from cosmic rays while providing no impediment to WIMPs themselves, spurious events will still exceed genuine ones by a large factor.

The UK Dark Matter Consortium (a joint venture by Imperial College, the University of Sheffield and the Rutherford Appleton Laboratory) operates WIMP detectors at the bottom of Boulby Mine in Yorkshire. Their sodium iodide and xenon detectors are designed to detect light emission from WIMP-induced events. Similar types of detector are operated by the Italian-Chinese DAMA (DArk MAtter)

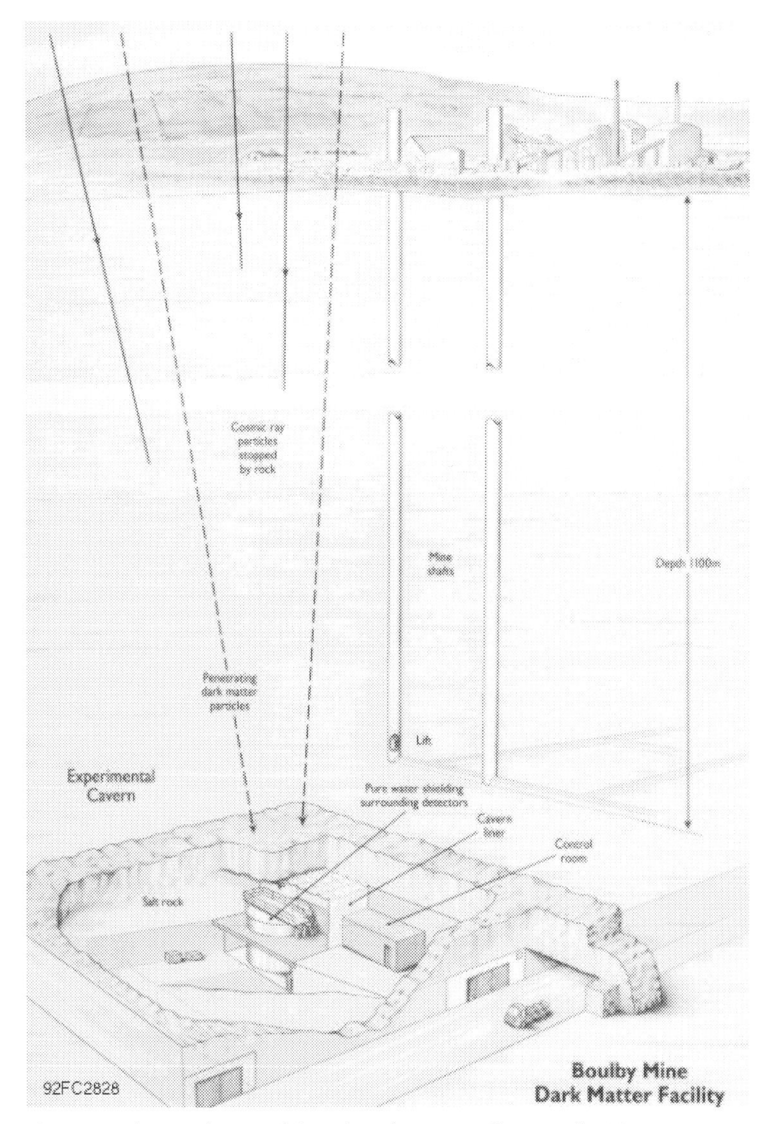

Figure 5. Schematic diagram of the UK's Dark Matter Facility at Boulby Mine. [Courtesy UK Dark Matter Collaboration]

team at the Gran Sasso underground laboratory. In the USA, the Stanford-based CDMS (Cryogenic Dark Matter Search) uses cooled germanium detectors which are capable of measuring both the heat and the charge generated by a recoiling nucleus and so, in principle, distinguishing between WIMP-induced and neutron-induced events.

At the time of writing (January 2001) there had been no clear-cut, undisputed detection of a WIMP. However, the DAMA team has claimed that a small seasonal variation in the total event rate in their detector provides statistically significant evidence for a population of WIMPs in our Galaxy. The argument is as follows. If the Galaxy has a halo populated with WIMPs then, because the Sun is orbiting around the galactic centre at a speed of around 225 km/s, it will appear to us as if a WIMP 'wind' is blowing past the Solar System. In addition, because the Earth is orbiting around the Sun, more WIMPs will be encountered when the Earth is heading in the same direction as the Sun (in June) as when it is heading in the opposite direction (in December). This should result in a periodic 10 per cent modulation of the WIMP event rate. From their analysis of four years' data, the DAMA team claim that there is a statistically significant 1 per cent modulation in the total event rate (which would include background events as well as WIMPs) which peaks on 2 June each year. They contend that their data is consistent with the existence of a galactic halo consisting of WIMPs with masses in the region of 60 proton masses.

Other researchers are sceptical. Some contend that a 1 per cent effect is not statistically significant and have suggested that the perceived fluctuations could be due to some kind of change in the background signal. The CDMS team, which by early 2000 had registered 13 hits which mimicked WIMP events, concluded that all of their events were almost certainly caused by stray neutrons rather than WIMPs. On the basis of their data, and the sensitivity of their detectors, they are confident that the DAMA results do not reveal the presence of WIMPs.

The controversy is likely to continue until more definitive data become available and until such time as researchers can identify individual events that have the clear-cut WIMP signature.

DARK MATTER IS NOT ENOUGH

Several strands of evidence suggest that the combined mean density of baryonic and non-baryonic matter is less than the critical density. In particular, because measurements of the internal dynamics of galaxy clusters give the total mass of the cluster (baryonic plus non-baryonic matter), and the total luminosity of the cluster gives a measure of the amount of baryonic matter it contains, it is possible to calculate the ratio of baryonic mass to total mass (the baryon fraction) in clusters. If the baryon fraction in the Universe as a whole is similar to that in clusters, then the overall matter density is about 5–7 times greater than that of baryonic matter.

When this information is combined with Big Bang nucleosynthesis arguments – which imply that baryons (luminous and dark) provide only about 3–5 per cent of the critical density – the results imply that matter as a whole (baryonic and non-baryonic) has a mean density equivalent to about one-third of the critical density.

NEW CLUES FROM THE MICROWAVE BACKGROUND

Detailed maps of the cosmic microwave background obtained by two balloon-borne experiments – BOOMERANG and MAXIMA – were published in the spring of the year 2000. Both sets of results reveal small-scale hotter and cooler patches in the microwave background. The predominant angular size of the strongest temperature variation (differences in temperature of about 50 millionths of a degree) were just under one degree (about twice the apparent size of the Moon in the sky). The sizes of these variations match closely what is predicted by the flat-space inflationary Universe model.

In the very early Universe, small differences in density between different regions of space would have arisen. In a region of enhanced density, gravity would cause particles of matter to fall inwards, but their inward motion would be halted and reversed by collisions with energetic photons. Clumps of plasma would bounce in and out, smaller clumps oscillating faster than larger ones. When the temperature everywhere in the Universe dropped to about 4000 K, some 300,000 years after the start of the expansion, atomic nuclei would have been able to

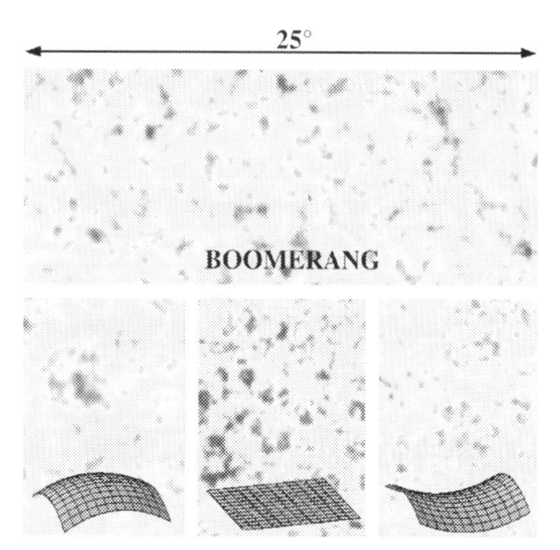

Figure 6. By observing the characteristic size of warmer and cooler spots in the BOOMERANG images, the geometry of space can be determined. They seem to indicate that space is very nearly 'flat', i.e. we live in a Universe in which standard high school geometry applies. [Courtesy BOOMERANG Collaboration and University of California Santa Barbara]

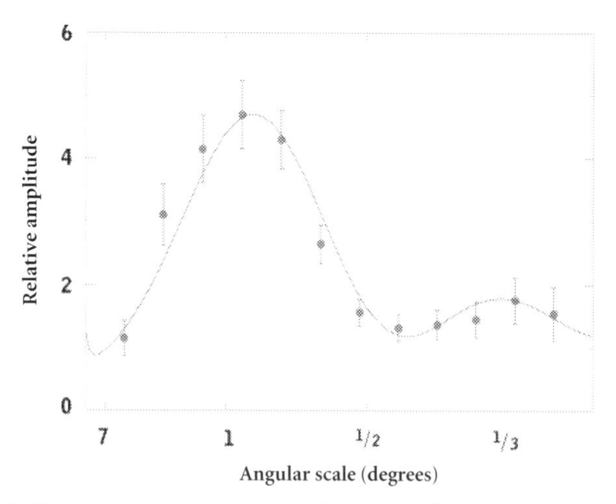

Figure 7. The power spectrum of sound waves in the primordial plasma from BOOMERANG data. These sound waves caused the temperature variations in the early Universe which are seen in the BOOMERANG images. [Courtesy BOOMERANG Collaboration and University of California Santa Barbara]

capture electrons to make complete atoms for the first time in the history of the Universe, a process that is called recombination. Space then became transparent, and the radiation content of the Universe was able to travel freely through the expanding volume of space with little chance of colliding and interacting with particles of matter. Diluted, stretched in wavelength and cooled by the expansion of space that has taken place since that time, this relic radiation from the Big Bang is visible now as a background of microwave radiation across the whole sky – the cosmic microwave background.

The cosmic microwave background radiation contains, in the form of hotter and cooler patches, the imprint of regions of marginally higher and lower density that existed at the time when that radiation last interacted with matter, when the Universe was about 300,000 years old. The temperature variations arise because light loses energy and is shifted to a marginally longer wavelength (which corresponds to a lower observed temperature) as it 'climbs out of' regions of enhanced density. Because the drag exerted by radiation on matter reduces the amplitude of each successive 'bounce', the strongest features correspond to patches of plasma which were at maximum compression, halfway through their first cycle of oscillation, at the time when the decoupling of matter and radiation took place. Inflationary theory makes very firm predictions about the angular sizes which these temperature fluctuations should have. If space is flat, the angular size of the dominant patches should be about 1 degree. In a closed, positively curved, space, the bending of light rays as they travel through space would cause the patches to appear larger, whereas in an open, negatively curved space, the patches would appear smaller.

The maps contain temperature fluctuations on many different scales. Using mathematical techniques to extract signals of different frequencies – corresponding to different angular scales – cosmologists can produce a 'power spectrum' which plots the amplitude ('height') of the fluctuations against their angular sizes. The angular power spectrum displays a series of peaks and troughs. The strongest peak in the power spectrum of the BOOMERANG and MAXIMA data corresponds to features on a scale of just under one degree, and it is this peak that provides strong evidence for a flat Universe.

The power spectrum ought to contain further, less prominent, peaks at angular scales of about a half and one-third of the angular size of the principal peak. The strength of the second peak will be influenced by the relative proportions of baryonic and dark matter in the Universe. Because oscillations of baryonic matter would have been heavily damped down by interactions with photons, whereas oscillations of CDM would not, the second peak would be higher if the Universe contains a large proportion of CDM than if CDM were a minor constituent. From the data so far, the second peak seems to be much weaker than would be expected in a Universe in which CDM outweighs baryonic matter by a factor of 5 or 10. Some have argued that the problem would vanish if the proportion of baryonic matter in the Universe were increased by a factor of between 1.5 and 2; but that seems to be in conflict with the Big Bang nucleosynthesis data. Alternatively, perhaps cold dark-matter particles interact with themselves in ways that could damp down the second peak. Using a modified theory of gravity called MOND (MOdified Newtonian Dynamics), conceived in 1983 by Moti Milgrom of the Weizmann Institute as an alternative to having to invoke dark matter, Stacy McGaugh, of the University of Maryland, contends that the observed power spectrum can be matched to a Universe which contains baryonic matter in the proportions given by Big Bang nucleosynthesis, and that CDM can be rejected altogether.

More data with greater sensitivity and higher resolution will be needed before any firm conclusions can be drawn about the proportion of CDM in the Universe. Although under fire from several quarters, non-baryonic cold dark matter remains for the moment an important potential component of the Universe. Whether or not the matter content of the Universe is dominated by CDM, the BOOMERANG and MAXIMA data imply that space is flat and that some commodity other than dark matter must make up the 'missing' two-thirds of the mass-energy density of the Universe.

DARK ENERGY, LAMBDA AND THE ACCELERATING UNIVERSE

In the mid-1990s, faced with fact that the observational evidence pointed towards a Universe with a matter density only one-third of what is needed to make the Universe flat, whereas the favoured inflation-CDM scenario requires a flat space with a mass-energy density equal to the critical density (a view that subsequently received strong observational support from the BOOMERANG/MAXIMA data), several groups of cosmologists suggested that the Universe might contain an extra ingredient – dark energy. Because energy affects the overall density and curvature of the Universe, dark energy – in the form of particles or fields – could provide the remaining two-thirds of the total mass-energy of the Universe.

One possible form of dark energy is the cosmological constant, lambda (Λ), an extra term which Einstein added to his general relativistic cosmological equations in order to account for a static Universe. At that time, which was before Hubble had shown that the Universe is expanding, it was generally believed that the Universe was static and that galaxies remained always at fixed distances. The cosmological constant was a property of space – a form of cosmic repulsion – which opposed gravity at very large distances. By choosing precisely the right value of lambda, Einstein could ensure that the gravitational attraction and cosmic repulsion exactly balanced and kept the galaxies at fixed distances from each other. Clearly such a balance is inherently unstable. If galaxies were nudged closer together, gravity would dominate, and they would begin to fall together, whereas if they were nudged a little further apart, cosmic repulsion would gain the upper hand and galaxies would begin to accelerate away from each other.

When Hubble discovered that the Universe is expanding, Einstein called lambda his 'greatest blunder' for, had he not introduced lambda, his equations would have predicted the expansion of the Universe before Hubble had discovered it. Since then, and up until very recently, cosmologists have usually assumed that the value of lambda is zero, since its existence appeared to be unnecessary to explain the observed properties of the expanding Universe. In the past few years, that situation has changed.

A finite value of the cosmological constant implies that space itself (the vacuum) contains a tiny constant energy density (vacuum energy). This would affect the curvature and density of the Universe and could make up the shortfall in the mass-energy that is needed to make the Universe flat. For a few years, in the late 1990s, many cosmologists worked on the assumption that about one-third of the mass-energy density was provided by matter (predominantly cold dark matter) and two-thirds by lambda.

AN ACCELERATING UNIVERSE?

In order to act as a form of cosmic repulsion, the cosmological constant has to have negative pressure. If the pressure is sufficiently negative, it will overwhelm gravity on the large scale and cause the expansion of the Universe to accelerate (rather than gently decelerating, as would be the case if it were dominated by matter).

In 1998, two research projects – the Supernova Cosmology Project and the High-redshift Supernova Search – published startling results which seem to indicate that the Universe is indeed in a state of accelerating expansion at this time. Both projects have been concentrating on measuring the apparent brightnesses of Type Ia supernovae in distant galaxies. Because supernovae of this kind are highly luminous, reach similar peak magnitudes, and their brightness rises and falls in a characteristic way that can be related to their peak luminosities, they provide particularly good indicators for measuring the distances of the remote galaxies within which they are embedded.

When the apparent brightnesses and redshifts of the observed supernovae were plotted on a diagram (a Hubble diagram) it emerged that the more distant supernovae were consistently fainter (and hence further away) than would be the case if the Universe were expanding at a steady rate, or if the rate of expansion were slowing down. The results appear to indicate that the Universe is expanding at an accelerating rate. Although it is too early to be certain that the results have been interpreted correctly, at the very least, they show that the Universe is not decelerating in the way that would be expected if it consisted of matter alone.

IS LAMBDA THE DARK ENERGY?

On the face of it, the supernova results appear to be consistent with a Universe with a finite (though tiny) cosmological constant, which is now in an accelerating phase of expansion. However, physicists and cosmologists are puzzled about why lambda should have so tiny a value. Because it is a constant, its value would have been utterly microscopic compared to the energy density of matter and radiation in the very early Universe. Why should it have started out with so tiny a value,

Figure 8. Hubble Space Telescope image of the edge-on spiral galaxy, NGC 4013. To account for the way in which spiral galaxies such as this rotate, they must contain 5–10 times as much dark matter as luminous matter. [Courtesy NASA and the Hubble Heritage Team (STScI/AURA)]

and why should it have become the dominant form of energy at this particular time, 15 billion years after the beginning of the expansion? Faced with these questions, a number of cosmologists and physicists are uneasy with the concept that vacuum energy density is the dominant constituent of the present-day Universe.

QUINTESSENCE

Alternative forms of dark energy – known as quintessence – were proposed in 1998 by Robert R. Caldwell (Dartmouth College, New Hampshire), Rahul Dave (University of Pennsylvania) and Paul J. Steinhardt (Princeton). There is a nice classical touch to the choice of name, 'quintessence' being the fifth element (after earth, water, air and fire) which the ancient Greek philosophers regarded as the unchanging pure element of the heavens. Key features of quintessence are that it should fill all space, not clump together like baryonic or cold dark matter, and that it should possess negative pressure. If the negative pressure is high enough, quintessence, like vacuum energy density, will drive the Universe into an accelerated phase of expansion.

The energy density of quintessence would have to have been feeble compared to the radiation and matter densities in the very early Universe. Only recently would the energy densities of matter and radiation have dropped below the energy density of quintessence and the Universe have entered a phase of accelerating expansion. One of the problems faced by any kind of dark energy (vacuum energy density or quintessence) is, why has it become the dominant component of the Universe at this particular time, or in the relatively recent past?

Of particular interest, therefore, are forms of quintessence which evolve with time. Quintessence may take the form of a 'tracker field', a field of a kind that is permitted by certain theories that attempt to unify the fundamental forces. In the very early Universe, the energy density of the tracker field would be small compared to the energy density of radiation and matter. Initially, the Universe was dominated by energetic radiation, but as it expanded and cooled, the radiation density declined more rapidly than the matter density. When the radiation density dropped below the matter density and the Universe switched from being radiation-dominated to being matter-dominated, its rate of expansion changed, and structure (galaxies and clusters) began to form.

Unlike vacuum energy density, tracker quintessence would change with time. It would mimic the decline in the density of radiation and matter, but in order to become the dominant form of energy, it would have to break that connection at some time in the history of the Universe. One particular form of tracker quintessence, known as k-essence (kinetic energy-driven quintessence), proposed in 2000 by

Armendariz, Mukhanov and Steinhardt, may have the desired properties. Its proponents suggest that k-essence would initially mimic the decline of radiation density but, when the transition from radiation to matter domination occurred, instead of tracking the matter density, it would 'freeze out' to a near constant value. Some time later, the matter density would drop below the k-essence density. Thereafter, k-essence would become the dominant component of the Universe and would drive it into a phase of accelerating expansion. Because the onset of the 'freezing out' of the tracker field and the formation of structure in the Universe were both triggered by the onset of matter domination, they argue that it is not surprising that dark energy should have become dominant in the relatively recent past.

Currently, theoreticians are exploring a wide range of possible quintessence models and are looking for ways to test these models against observations.

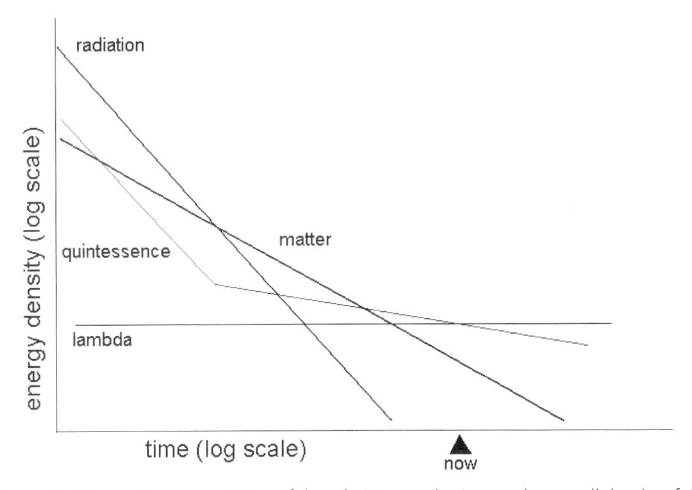

Figure 9. Schematic representation of the relative contributions to the overall density of the Universe of radiation, matter, one possible form of quintessence (k-essence) and the cosmological constant (lambda). In this representation, dark energy (quintessence or lambda) has become the dominant constituent of the Universe relatively recently.

INVENTORY OF THE UNIVERSE

Cosmology is in an exciting phase. Observations appear to indicate that space is flat and that the Universe may be expanding at an accelerating rate. A whole range of observational evidence shows clearly that the Universe contains a great deal more dark matter than luminous matter and that – even though doubts have been expressed – most of the dark matter appears to be non-baryonic cold dark matter. The overall density of dark matter in the Universe appears to be no more than a third of the critical density that a flat-space inflationary Universe needs to have. Dark energy, therefore, appears to be the dominant constituent of the Universe at this time.

It looks as if the relative contributions to the overall density of the Universe contributed by its various different constituents may be something like this: luminous matter 0.5–1 per cent, neutrinos 0.3 per cent, baryonic dark matter 3–5 per cent, non-baryonic dark matter (WIMPs or other forms of cold dark matter) 20–30 per cent, and dark energy 65–70 per cent. Ground-based instruments such as The Degree Angular Scale Interferometer (DASI), ongoing balloon-borne experiments, and spacecraft such as NASA's Microwave Anisotropy Probe (MAP), launched in June 2001, and ESA's Planck Mission (scheduled for launching in 2007), should greatly improve the precision of these figures within the next few years.

One thing, however, appears certain. Visible matter provides only a tiny fraction of the total mass-energy content of the Universe. Whether it consists of baryons, WIMPs, dark energy or something else, most of its mass-energy is dark. The splendid panorama of stars, nebulae and galaxies revealed by astronomers' telescopes is little more than thin icing on a dark and ponderous cosmic cake.

Part III

Miscellaneous

Some Interesting Variable Stars

JOHN ISLES

The following stars are of interest for many reasons. Of course, the periods and ranges of many variables are not constant from one cycle to another. Finder charts are given on the pages following this list for those stars marked with an asterisk.

Star	RA		Declination		Range	Type	Period	Spectrum
	h	m	°	′			(days)	
R Andromedae	00	24.0	+38	35	5.8–14.9	Mira	409	S
W Andromedae	02	17.6	+44	18	6.7–14.6	Mira	396	S
U Antliae	10	35.2	−39	34	5–6	Irregular	—	C
Theta Apodis	14	05.3	−76	48	5–7	Semi-regular	119	M
R Aquarii	23	43.8	−15	17	5.8–12.4	Symbiotic	387	M+Pec
T Aquarii	20	49.9	−05	09	7.2–14.2	Mira	202	M
R Aquilae	19	06.4	+08	14	5.5–12.0	Mira	284	M
V Aquilae	19	04.4	−05	41	6.6–8.4	Semi-regular	353	C
Eta Aquilae	19	52.5	+01	00	3.5–4.4	Cepheid	7.2	F–G
U Arae	17	53.6	−51	41	7.7–14.1	Mira	225	M
R Arietis	02	16.1	+25	03	7.4–13.7	Mira	187	M
U Arietis	03	11.0	+14	48	7.2–15.2	Mira	371	M
R Aurigae	05	17.3	+53	35	6.7–13.9	Mira	458	M
Epsilon Aurigae	05	02.0	+43	49	2.9–3.8	Algol	9892	F+B
R Boötis	14	37.2	+26	44	6.2–13.1	Mira	223	M
W Boötis	14	43.4	+26	32	4.7–5.4	Semi-regular?	450?	M
X Camelopardalis	04	45.7	+75	06	7.4–14.2	Mira	144	K–M
R Cancri	08	16.6	+11	44	6.1–11.8	Mira	362	M
X Cancri	08	55.4	+17	14	5.6–7.5	Semi-regular	195?	C
*R Canis Majoris	07	19.5	−16	24	5.7–6.3	Algol	1.1	F
*VY Canis Majoris	07	23.0	−25	46	6.5–9.6	Unique	—	M
*FW Canis Majoris	07	24.7	−16	12	5.0–5.5	Gamma Cas	—	B
S Canis Minoris	07	32.7	+08	19	6.6–13.2	Mira	333	M
R Canum Ven.	13	49.0	+39	33	6.5–12.9	Mira	329	M
R Carinae	09	32.2	−62	47	3.9–10.5	Mira	309	M
S Carinae	10	09.4	−61	33	4.5–9.9	Mira	149	K–M
l Carinae	09	45.2	−62	30	3.3–4.2	Cepheid	35.5	F–K
Eta Carinae	10	45.1	−59	41	−0.8–7.9	Irregular	—	Pec

Star	RA		Declination		Range	Type	Period	Spectrum
	h	m	°	′			(days)	
R Cassiopeiae	23	58.4	+51	24	4.7–13.5	Mira	430	M
S Cassiopeiae	01	19.7	+72	37	7.9–16.1	Mira	612	S
W Cassiopeiae	00	54.9	+58	34	7.8–12.5	Mira	406	C
Gamma Cas.	00	56.7	+60	43	1.6–3.0	Gamma Cas.	—	B
Rho Cassiopeiae	23	54.4	+57	30	4.1–6.2	Semi-regular	—	F–K
R Centauri	14	16.6	–59	55	5.3–11.8	Mira	546	M
S Centauri	12	24.6	–49	26	7–8	Semi-regular	65	C
T Centauri	13	41.8	–33	36	5.5–9.0	Semi-regular	90	K–M
S Cephei	21	35.2	+78	37	7.4–12.9	Mira	487	C
T Cephei	21	09.5	+68	29	5.2–11.3	Mira	388	M
Delta Cephei	22	29.2	+58	25	3.5–4.4	Cepheid	5.4	F–G
Mu Cephei	21	43.5	+58	47	3.4–5.1	Semi-regular	730	M
U Ceti	02	33.7	–13	09	6.8–13.4	Mira	235	M
W Ceti	00	02.1	–14	41	7.1–14.8	Mira	351	S
*Omicron Ceti	02	19.3	–02	59	2.0–10.1	Mira	332	M
R Chamaeleontis	08	21.8	–76	21	7.5–14.2	Mira	335	M
T Columbae	05	19.3	–33	42	6.6–12.7	Mira	226	M
R Comae Ber.	12	04.3	+18	47	7.1–14.6	Mira	363	M
R Coronae Bor.	15	48.6	+28	09	5.7–14.8	R Coronae Bor.	—	C
S Coronae Bor.	15	21.4	+31	22	5.8–14.1	Mira	360	M
T Coronae Bor.	15	59.6	+25	55	2.0–10.8	Recurrent nova	—	M+Pec
V Coronae Bor.	15	49.5	+39	34	6.9–12.6	Mira	358	C
W Coronae Bor.	16	15.4	+37	48	7.8–14.3	Mira	238	M
R Corvi	12	19.6	–19	15	6.7–14.4	Mira	317	M
R Crucis	12	23.6	–61	38	6.4–7.2	Cepheid	5.8	F–G
R Cygni	19	36.8	+50	12	6.1–14.4	Mira	426	S
U Cygni	20	19.6	+47	54	5.9–12.1	Mira	463	C
W Cygni	21	36.0	+45	22	5.0–7.6	Semi-regular	131	M
RT Cygni	19	43.6	+48	47	6.0–13.1	Mira	190	M
SS Cygni	21	42.7	+43	35	7.7–12.4	Dwarf nova	50±	K+Pec
CH Cygni	19	24.5	+50	14	5.6–9.0	Symbiotic	—	M+B
Chi Cygni	19	50.6	+32	55	3.3–14.2	Mira	408	S
R Delphini	20	14.9	+09	05	7.6–13.8	Mira	285	M
U Delphini	20	45.5	+18	05	5.6–7.5	Semi-regular	110?	M
EU Delphini	20	37.9	+18	16	5.8–6.9	Semi-regular	60	M
Beta Doradus	05	33.6	–62	29	3.5–4.1	Cepheid	9.8	F–G
R Draconis	16	32.7	+66	45	6.7–13.2	Mira	246	M
T Eridani	03	55.2	–24	02	7.2–13.2	Mira	252	M
R Fornacis	02	29.3	–26	06	7.5–13.0	Mira	389	C
R Geminorum	07	07.4	+22	42	6.0–14.0	Mira	370	S

Star	RA		Declination		Range	Type	Period	Spectrum
	h	m	°	′			(days)	
U Geminorum	07	55.1	+22	00	8.2–14.9	Dwarf nova	105±	Pec+M
*Zeta Geminorum	07	04.1	+20	34	3.6–4.2	Cepheid	10.2	F–G
*Eta Geminorum	06	14.9	+22	30	3.2–3.9	Semi-regular	233	M
S Gruis	22	26.1	–48	26	6.0–15.0	Mira	402	M
S Herculis	16	51.9	+14	56	6.4–13.8	Mira	307	M
U Herculis	16	25.8	+18	54	6.4–13.4	Mira	406	M
Alpha Herculis	17	14.6	+14	23	2.7–4.0	Semi-regular	—	M
68, u Herculis	17	17.3	+33	06	4.7–5.4	Algol	2.1	B+B
R Horologii	02	53.9	–49	53	4.7–14.3	Mira	408	M
U Horologii	03	52.8	–45	50	6–14	Mira	348	M
R Hydrae	13	29.7	–23	17	3.5–10.9	Mira	389	M
U Hydrae	10	37.6	–13	23	4.3–6.5	Semi-regular	450?	C
VW Hydri	04	09.1	–71	18	8.4–14.4	Dwarf nova	27±	Pec
R Leonis	09	47.6	+11	26	4.4–11.3	Mira	310	M
R Leonis Minoris	09	45.6	+34	31	6.3–13.2	Mira	372	M
R Leporis	04	59.6	–14	48	5.5–11.7	Mira	427	C
Y Librae	15	11.7	–06	01	7.6–14.7	Mira	276	M
RS Librae	15	24.3	–22	55	7.0–13.0	Mira	218	M
Delta Librae	15	01.0	–08	31	4.9–5.9	Algol	2.3	A
R Lyncis	07	01.3	+55	20	7.2–14.3	Mira	379	S
R Lyrae	18	55.3	+43	57	3.9–5.0	Semi-regular	46?	M
RR Lyrae	19	25.5	+42	47	7.1–8.1	RR Lyrae	0.6	A–F
Beta Lyrae	18	50.1	+33	22	3.3–4.4	Eclipsing	12.9	B
U Microscopii	20	29.2	–40	25	7.0–14.4	Mira	334	M
*U Monocerotis	07	30.8	–09	47	5.9–7.8	RV Tauri	91	F–K
V Monocerotis	06	22.7	–02	12	6.0–13.9	Mira	340	M
R Normae	15	36.0	–49	30	6.5–13.9	Mira	508	M
T Normae	15	44.1	–54	59	6.2–13.6	Mira	241	M
R Octantis	05	26.1	–86	23	6.3–13.2	Mira	405	M
S Octantis	18	08.7	–86	48	7.2–14.0	Mira	259	M
V Ophiuchi	16	26.7	–12	26	7.3–11.6	Mira	297	C
X Ophiuchi	18	38.3	+08	50	5.9–9.2	Mira	329	M
RS Ophiuchi	17	50.2	–06	43	4.3–12.5	Recurrent nova	—	OB+M
U Orionis	05	55.8	+20	10	4.8–13.0	Mira	368	M
W Orionis	05	05.4	+01	11	5.9–7.7	Semi-regular	212	C
Alpha Orionis	05	55.2	+07	24	0.0–1.3	Semi-regular	2335	M
S Pavonis	19	55.2	–59	12	6.6–10.4	Semi-regular	381	M
Kappa Pavonis	18	56.9	–67	14	3.9–4.8	Cepheid	9.1	G
R Pegasi	23	06.8	+10	33	6.9–13.8	Mira	378	M
Beta Pegasi	23	03.8	+28	05	2.3–2.7	Irregular	—	M

Star	RA		Declination		Range	Type	Period	Spectrum
	h	m	°	′			(days)	
X Persei	03	55.4	+31	03	6.0–7.0	Gamma Cas.	—	09.5
Beta Persei	03	08.2	+40	57	2.1–3.4	Algol	2.9	B
Rho Persei	03	05.2	+38	50	3.3–4.0	Semi-regular	50?	M
Zeta Phoenicis	01	08.4	–55	15	3.9–4.4	Algol	1.7	B+B
R Pictoris	04	46.2	–49	15	6.4–10.1	Semi-regular	171	M
*RS Puppis	08	13.1	–34	35	6.5–7.7	Cepheid	41.4	F-G
L^2 Puppis	07	13.5	–44	39	2.6–6.2	Semi-regular	141	M
T Pyxidis	09	04.7	–32	23	6.5–15.3	Recurrent nova	7000±	Pec
U Sagittae	19	18.8	+19	37	6.5–9.3	Algol	3.4	B+G
WZ Sagittae	20	07.6	+17	42	7.0–15.5	Dwarf nova	1900±	A
R Sagittarii	19	16.7	–19	18	6.7–12.8	Mira	270	M
RR Sagittarii	19	55.9	–29	11	5.4–14.0	Mira	336	M
RT Sagittarii	20	17.7	–39	07	6.0–14.1	Mira	306	M
RU Sagittarii	19	58.7	–41	51	6.0–13.8	Mira	240	M
RY Sagittarii	19	16.5	–33	31	5.8–14.0	R Coronae Bor.	—	G
RR Scorpii	16	56.6	–30	35	5.0–12.4	Mira	281	M
RS Scorpii	16	55.6	–45	06	6.2–13.0	Mira	320	M
RT Scorpii	17	03.5	–36	55	7.0–15.2	Mira	449	S
S Sculptoris	00	15.4	–32	03	5.5–13.6	Mira	363	M
R Scuti	18	47.5	–05	42	4.2–8.6	RV Tauri	146	G–K
R Serpentis	15	50.7	+15	08	5.2–14.4	Mira	356	M
S Serpentis	15	21.7	+14	19	7.0–14.1	Mira	372	M
T Tauri	04	22.0	+19	32	9.3–13.5	T Tauri	—	F–K
SU Tauri	05	49.1	+19	04	9.1–16.9	R Coronae Bor.	—	G
Lambda Tauri	04	00.7	+12	29	3.4–3.9	Algol	4.0	B+A
R Trianguli	02	37.0	+34	16	5.4–12.6	Mira	267	M
R Ursae Majoris	10	44.6	+68	47	6.5–13.7	Mira	302	M
T Ursae Majoris	12	36.4	+59	29	6.6–13.5	Mira	257	M
U Ursae Minoris	14	17.3	+66	48	7.1–13.0	Mira	331	M
R Virginis	12	38.5	+06	59	6.1–12.1	Mira	146	M
S Virginis	13	33.0	–07	12	6.3–13.2	Mira	375	M
SS Virginis	12	25.3	+00	48	6.0–9.6	Semi-regular	364	C
R Vulpeculae	21	04.4	+23	49	7.0–14.3	Mira	137	M
Z Vulpeculae	19	21.7	+25	34	7.3–8.9	Algol	2.5	B+A

R and FW Canis Majoris

Comparison stars:

A = 4.96
B = 5.45
C = 5.46
D = 5.78
E = 6.05
F = 6.09
G = 6.6
H = 6.77

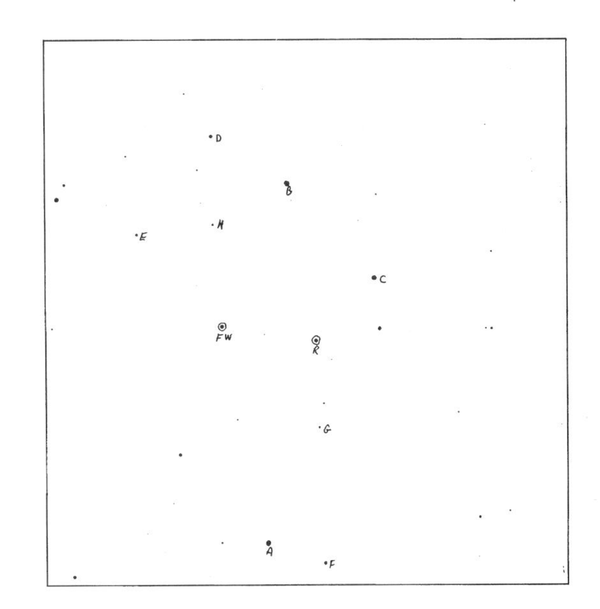

VY Canis Majoris

Comparison stars:

C = 7.0
D = 7.1
E = 8.1
F = 8.4
G = 8.8
H = 9.4

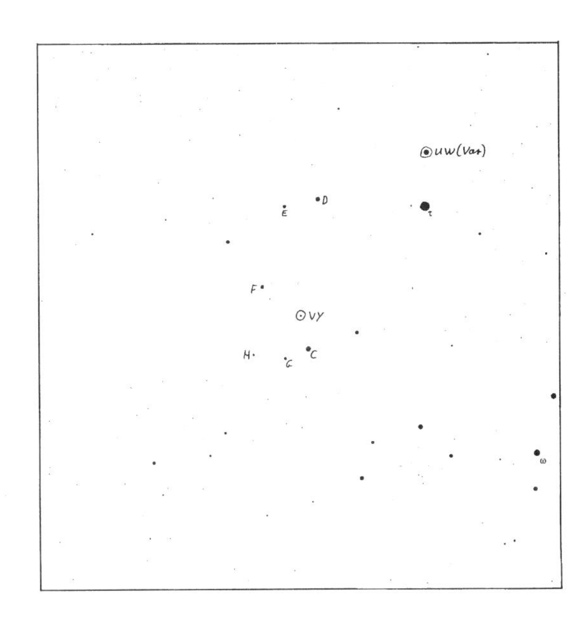

Omicron Ceti (Mira)

Comparison stars:

Alpha (α)	= 2.52 (off map)
Gamma (γ)	= 3.46
Delta (δ)	= 4.06
Nu (ν)	= 4.87
N	= 5.34
P	= 5.41
R	= 6.00
S	= 6.32
T	= 6.49
U	= 7.19
W	= 8.06
X	= 8.42
y	= 9.00
z	= 9.33

Eta and Zeta Geminorum

Comparison stars:

Epsilon (ε) Gem	= 2.98
Zeta (ζ) Tau	= 3.03
Xi (χ) Gem	= 3.34
Lambda (λ) Gem	= 3.59
Nu (ν) Gem	= 4.14
1 Gem	= 4.15

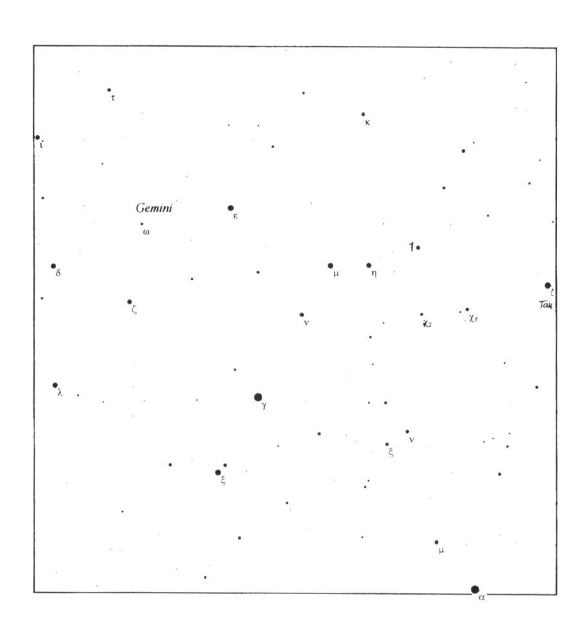

U Monocerotis

Comparison stars:

C	= 5.72
D	= 5.85
E	= 6.00
F	= 6.62
G	= 6.97
H	= 7.51
K	= 7.81
L	= 8.03

RS Puppis

Comparison stars:

A	= 6.4
B	= 6.4
C	= 7.0
D	= 7.4
E	= 7.6
F	= 8.2
G	= 8.3

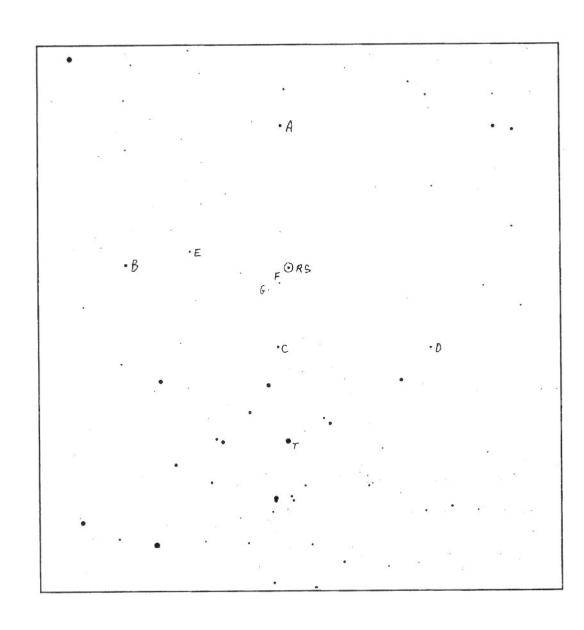

Mira Stars: Maxima, 2002

JOHN ISLES

Below are the predicted dates of maxima for Mira stars that reach magnitude 7.5 or brighter at an average maximum. Individual maxima can in some cases be brighter or fainter than average by a magnitude or more, and all dates are only approximate. The positions, extreme ranges and mean periods of these stars can be found in the preceding list of interesting variable stars.

Star	Mean magnitude at maximum	Dates of maxima
R Andromedae	6.9	Nov 15
W Andromedae	7.4	Dec 2
R Aquarii	6.5	Aug 1
R Aquilae	6.1	Apr 14
R Boötis	7.2	Mar 17, Oct 26
R Cancri	6.8	Oct 11
S Canis Minoris	7.5	Sep 24
R Carinae	4.6	Jul 11
S Carinae	5.7	Jan 20, Jun 19, Nov 15
R Centauri	5.8	Mar 10
T Cephei	6.0	Aug 31
U Ceti	7.5	Jul 13
Omicron Ceti	3.4	Jul 31
T Columbae	7.5	Mar 17, Oct 29
S Coronae Borealis	7.3	Oct 15
V Coronae Borealis	7.5	Apr 13
R Corvi	7.5	Jan 29, Dec 12
U Cygni	7.2	Jun 3
RT Cygni	7.3	Apr 10, Oct 17
Chi Cygni	5.2	Feb 8
R Geminorum	7.1	Nov 22
U Herculis	7.5	Feb 8

Star	Mean magnitude at maximum	Dates of maxima
R Horologii	6.0	Apr 22
R Hydrae	4.5	Jul 18
R Leonis	5.8	Jan 7, Nov 13
R Leonis Minoris	7.1	Oct 18
R Leporis	6.8	Sep 22
RS Librae	7.5	May 12, Dec 16
V Monocerotis	7.0	Feb 21
R Normae	7.2	Jan 28
T Normae	7.4	Mar 27, Nov 23
V Ophiuchi	7.5	Aug 28
X Ophiuchi	6.8	Jun 4
U Orionis	6.3	Dec 7
R Sagittarii	7.3	Aug 6
RR Sagittarii	6.8	Aug 3
RT Sagittarii	7.0	Sep 9
RU Sagittarii	7.2	Jul 14
RR Scorpii	5.9	Apr 18
RS Scorpii	7.0	Mar 14
S Sculptoris	6.7	Dec 10
R Serpentis	6.9	Jan 15
R Trianguli	6.2	Jun 1
R Ursae Majoris	7.5	Aug 7
R Virginis	6.9	Jan 3, May 29, Oct 21
S Virginis	7.0	Dec 22

Some Interesting Double Stars

R. W. ARGYLE

The positions, angles and separations given below correspond to epoch 2002.0.

No	RA		Declination		Star	Magnitudes	Separation	PA	Comments
	h	m	°	′			″	°	
1	00	31.5	−62	58	β Tuc	4.4, 4.8	27.1	169	LCL 119 Both again difficult doubles.
2	00	49.1	+57	49	η Cas	3.4, 7.5	12.9	317	STF 60 Easy. Creamy, bluish.
3	00	55.0	+23	38	36 And	6.0, 6.4	0.9	312	STF 73 P = 168 years. Both yellow. Slowly opening.
4	01	13.7	+07	35	ζ Psc	5.6, 6.5	23.1	63	Yellow, reddish-white.
5	01	39.8	−56	12	p Eri	5.8, 5.8	11.6	190	Period = 483 years.
6	01	53.5	+19	18	γ Ari	4.8, 4.8	7.5	1	Very easy. Both white.
7	02	02.0	+02	46	α Psc	4.2, 5.1	1.8	270	Binary, period = 933 years.
8	02	03.9	+42	20	γ And	2.3, 5.0	9.6	63	Yellow, blue. Relatively fixed.
					γ2 And	5.1, 6.3	0.4	102	BC. Needs 30 cm. Closing.
9	02	29.1	+67	24	ι Cas AB	4.9, 6.9	2.6	231	AB is long period binary. P = 630 years.
					ι Cas AC	4.9, 8.4	7.2	118	
10	02	33.8	−28	14	ω For	5.0, 7.7	10.8	245	Common proper motion.
11	02	43.3	+03	14	γ Cet	3.5, 7.3	2.6	298	Not too easy.
12	02	58.3	−40	18	θ Eri	3.4, 4.5	8.3	90	Both white.
13	02	59.2	+21	20	ε Ari	5.2, 5.5	1.5	208	Binary. Little motion. Both white.

No	RA h	RA m	Declination °	Declination ′	Star	Magnitudes	Separation ″	PA °	Comments
14	03	00.9	+52	21	Σ331 Per	5.3, 6.7	12.0	85	Fixed.
15	03	12.1	−28	59	α For	4.0, 7.0	5.0	299	P = 269 years. B variable?
16	03	48.6	−37	37	f Eri	4.8, 5.3	8.2	215	Pale yellow. Fixed.
17	03	54.3	−02	57	32 Eri	4.8 ,6.1	6.9	348	Fixed.
18	04	32.0	+53	55	1 Cam	5.7, 6.8	10.3	308	Fixed.
19	04	50.9	−53	28	ι Pic	5.6, 6.4	12.4	58	Good object for small apertures. Fixed.
20	05	13.2	−12	56	κ Lep	4.5, 7.4	2.2	357	Visible in 7.5 cm.
21	05	14.5	−08	12	β Ori	0.1, 6.8	9.5	204	Companion once thought to be close double.
22	05	21.8	−24	46	41 Lep	5.4, 6.6	3.4	93	Deep yellow pair in a rich field.
23	05	24.5	−02	24	η Ori	3.8, 4.8	1.7	78	Slow moving binary.
24	05	35.1	+09	56	λ Ori	3.6, 5.5	4.3	44	Fixed.
25	05	35.3	−05	23	θ Ori AB	6.7, 7.9	8.6	32	Trapezium in M42.
					θ Ori CD	5.1, 6.7	13.4	61	
26	05	38.7	−02	36	σ Ori AC	4.0, 10.3	11.4	238	Quintuple. A is a close double.
					σ Ori ED	6.5, 7.5	30.1	231	
27	05	40.7	−01	57	ζ Ori	1.9, 4.0	2.4	164	Can be split in 7.5 cm. Long period binary.
28	06	14.9	+22	30	η Gem	var, 6.5	1.6	256	Well seen with 20 cm. Primary orange.
29	06	46.2	+59	27	12 Lyn AB	5.4, 6.0	1.7	69	AB is binary, P = 706 years.
					12 Lyn AC	5.4, 7.3	8.7	309	
30	07	08.7	−70	30	γ Vol	3.9, 5.8	14.1	298	Very slow binary.
31	07	16.6	−23	19	h3945 CMa	4.8, 6.8	26.8	51	Contrasting colours.
32	07	20.1	+21	59	δ Gem	3.5, 8.2	5.7	225	Not too easy. Yellow, pale blue.
33	07	34.6	+31	53	α Gem	1.9, 2.9	4.0	63	Widening. Easy with 7.5 cm.
34	07	38.8	−26	48	κ Pup	4.5, 4.7	9.8	318	Both white.
35	08	12.2	+17	39	ζ Cnc AB	5.6, 6.0	0.9	71	P (AB) = 60 years. C has infra-red companion.
					ζ Cnc AB–C	5.0, 6.2	5.9	72	

No	RA		Declination		Star	Magnitudes	Separation	PA	Comments
	h	m	°	′			″	°	
36	08	44.7	−54	43	δ Vel	2.1, 5.1	1.0	339	Difficult close pair. Period 142 years.
37	08	46.8	+06	25	ε Hyd	3.3, 6.8	2.9	299	PA slowly increasing. A is a very close pair.
38	09	18.8	+36	48	38 Lyn	3.9, 6.6	2.8	230	Almost fixed.
39	09	47.1	−65	04	μ Car	3.1, 6.1	5.0	128	Fixed. Fine in small telescopes.
40	10	20.0	+19	50	γ Leo	2.2, 3.5	4.4	125	Binary, period = 619 years. Both orange.
41	10	32.0	−45	04	s Vel	6.2, 6.5	13.5	218	Fixed.
42	10	46.8	−49	26	μ Vel	2.7, 6.4	2.5	54	P = 138 years. Near widest separation.
43	10	55.6	+24	45	54 Leo	4.5, 6.3	6.6	111	Slowly widening. Pale yellow and white.
44	11	18.2	+31	32	ξ UMa	4.3, 4.8	1.8	263	Binary, 60 years. Opening. Needs 7.5 cm.
45	11	21.0	−54	29	π Cen	4.3, 5.0	0.3	147	Binary, 38.7 years. Very close. Needs 35 cm.
46	11	23.9	+10	32	ι Leo	4.0, 6.7	1.8	109	Binary, period = 186 years.
47	11	32.3	−29	16	N Hya	5.8, 5.9	9.5	210	Fixed.
48	12	14.0	−45	43	D Cen	5.6, 6.8	2.8	243	Orange and white. Closing.
49	12	26.6	−63	06	α Cru	1.4, 1.9	4.0	112	Third star in a low power field.
50	12	41.5	−48	58	γ Cen	2.9, 2.9	0.9	346	Period = 84 years. Closing. Both yellow.
51	12	41.7	−01	27	γ Vir	3.5, 3.5	1.1	248	Binary. P = 168 years. Closing quickly.
52	12	46.3	−68	06	β Mus	3.7, 4.0	1.3	44	Both white. Closing slowly. P = 383 years.
53	12	54.6	−57	11	μ Cru	4.3, 5.3	34.9	17	Fixed. Both white.
54	12	56.0	+38	19	α CVn	2.9, 5.5	19.3	229	Easy. Yellow, bluish.

No	RA		Declination		Star	Magnitudes	Separation	PA	Comments
	h	m	°	′			″	°	
55	13	22.6	−60	59	J Cen	4.6, 6.5	60.0	343	Fixed. A is a close pair.
56	13	24.0	+54	56	ζ UMa	2.3, 4.0	14.4	152	Very easy. Naked eye pair with Alcor.
57	13	51.8	−33	00	3 Cen	4.5, 6.0	7.9	106	Both white. Closing slowly.
58	14	39.6	−60	50	α Cen	0.0, 1.2	14.1	222	Finest pair in the sky. P = 80 years. Closing.
59	14	41.1	+13	44	ζ Boo	4.5, 4.6	0.8	299	Both white. Closing – highly inclined orbit.
60	14	45.0	+27	04	ε Boo	2.5, 4.9	2.9	345	Yellow, blue. Fine pair.
61	14	46.0	−25	27	54 Hya	5.1, 7.1	8.3	122	Closing slowly.
62	14	49.3	−14	09	μ Lib	5.8, 6.7	1.9	2	Becoming wider. Fine in 7.5 cm.
63	14	51.4	+19	06	ξ Boo	4.7, 7.0	6.6	316	Fine contrast. Easy.
64	15	03.8	+47	39	44 Boo	5.3, 6.2	2.0	55	Period = 246 years.
65	15	05.1	−47	03	π Lup	4.6, 4.7	1.7	67	Widening.
66	15	18.5	−47	53	μ Lup AB	5.1, 5.2	1.0	128	AB closing?
					μ Lup AC	4.4, 7.2	23.2	129	AC almost fixed.
67	15	23.4	−59	19	γ Cir	5.1, 5.5	0.8	356	Closing. Needs 20 cm. Long period binary.
68	15	32.0	+32	17	η Cr B	5.6, 5.9	0.6	77	Both yellow. P = 41 years. Closing.
69	15	34.8	+10	33	δ Ser	4.2, 5.2	4.3	176	Long period binary.
70	15	35.1	−41	10	γ Lup	3.5, 3.6	0.8	278	Binary. Period = 190 years. Needs 20 cm.
71	15	56.9	−33	58	ξ Lup	5.3, 5.8	10.2	49	Fixed.
72	16	14.7	+33	52	σ CrB	5.6, 6.6	7.0	236	Long period binary. Both white.
73	16	29.4	−26	26	α Sco	1.2, 5.4	2.6	274	Red, green. Difficult from mid-northern latitudes.
74	16	30.9	+01	59	λ Oph	4.2, 5.2	1.5	31	P = 129 years. Fairly difficult in small apertures.
75	16	41.3	+31	36	ζ Her	2.9, 5.5	0.5	298	Period 34 years. Closing rapidly. Needs 30 cm.
76	17	05.3	+54	28	μ Dra	5.7, 5.7	2.2	15	Period 672 years.

No	RA h	RA m	Declination °	Declination ′	Star	Magnitudes	Separation ″	PA °	Comments
77	17	14.6	+14	24	α Her	var, 5.4	4.6	104	Red, green. Long period binary.
78	17	15.3	−26	35	36 Oph	5.1, 5.1	4.9	326	Period = 471 years.
79	17	23.7	+37	08	ρ Her	4.6, 5.6	4.1	318	Slowly widening.
80	18	01.5	+21	36	95 Her	5.0, 5.1	6.4	257	Colours thought variable in C19.
81	18	05.5	+02	30	70 Oph	4.2, 6.0	4.2	144	Opening. Easy in 7.5 cm.
82	18	06.8	−43	25	h5014 CrA	5.7, 5.7	1.7	6	Period = 450 years. Needs 10 cm.
83	18	35.9	+16	58	OΣ358 Her	6.8, 7.0	1.6	154	Period = 380 years.
84	18	44.3	+39	40	ε^1 Lyr	5.0, 6.1	2.6	350	Quadruple system with ε^2. Both pairs visible in 7.5 cm
85	18	44.3	+39	40	ε^2 Lyr	5.2, 5.5	2.3	82	
86	18	56.2	+04	12	θ Ser	4.5, 5.4	22.4	104	Fixed. Very easy.
87	19	06.4	−37	04	γ CrA	4.8, 5.1	1.3	53	Beautiful pair. Period = 122 years.
88	19	30.7	+27	58	β Cyg AB	3.1, 5.1	34.3	54	Glorious. Yellow, blue-greenish.
					β Cyg Aa	3.1, 4.0	0.4	122	Aa. Discovered in 1976. Period = 97 years.
89	19	45.0	+45	08	δ Cyg	2.9, 6.3	2.6	224	Slowly widening. Period = 780 years.
90	19	48.2	+70	16	ε Dra	3.8, 7.4	3.2	17	Slow binary.
91	20	46.7	+16	07	γ Del	4.5, 5.5	9.2	265	Easy. Yellowish. Long period binary.
92	20	47.4	+36	29	λ Cyg	4.8, 6.1	0.9	10	Difficult binary in small apertures.
93	20	59.1	+04	18	ε Equ AB	6.0, 6.3	0.8	285	Fine triple. AB is closing.
					ε Equ AC	6.0, 7.1	10.3	66	
94	21	06.9	+38	45	61 Cyg	5.2, 6.0	30.7	150	Nearby binary. Both orange. Period = 722 years.

No	RA		Declin-ation		Star	Magni-tudes	Separa-tion	PA	Comments
	h	m	°	′			″	°	
95	21	19.9	−53	27	θ Ind	4.5, 7.0	6.8	271	Pale yellow and reddish. Long period binary.
96	21	44.1	+28	45	μ Cyg	4.8, 6.1	1.8	310	Period = 713 years.
97	22	03.8	+64	37	ξ Cep	4.4, 6.5	8.1	275	White and blue. Long period binary.
98	22	26.6	−16	45	53 Aqr	6.4, 6.6	1.6	11	Long period binary, approaching periastron.
99	22	28.8	−00	01	ζ Aqr	4.3, 4.5	2.0	182	Slowly widening.
100	23	59.4	+33	43	Σ3050 And	6.6, 6.6	2.0	329	Period = 350 years.

Some Interesting Nebulae, Clusters and Galaxies

Object	RA		Declina-tion		Remarks
	h	m	°	′	
M31 Andromedae	00	40.7	+41	05	Andromeda Galaxy, visible to naked eye.
H VIII 78 Cassiopeiae	00	41.3	+61	36	Fine cluster, between Gamma and Kappa Cassiopeiae.
M33 Trianguli	01	31.8	+30	28	Spiral. Difficult with small apertures.
H VI 33–4 Persei, C14	02	18.3	+56	59	Double Cluster; Sword-handle.
Δ142 Doradus	05	39.1	−69	09	Looped nebula round 30 Doradus. Naked-eye. In Large Magellanic Cloud.
M1 Tauri	05	32.3	+22	00	Crab Nebula, near Zeta Tauri.
M42 Orionis	05	33.4	−05	24	Orion Nebula. Contains the famous Trapezium, Theta Orionis.
M35 Geminorum	06	06.5	+24	21	Open cluster near Eta Geminorum.
H VII 2 Monocerotis, C50	06	30.7	+04	53	Open cluster, just visible to naked eye.
M41 Canis Majoris	06	45.5	−20	42	Open cluster, just visible to naked eye.
M47 Puppis	07	34.3	−14	22	Mag. 5.2. Loose cluster.
H IV 64 Puppis	07	39.6	−18	05	Bright planetary in rich neighbourhood.
M46 Puppis	07	39.5	−14	42	Open cluster.
M44 Cancri	08	38	+20	07	Praesepe. Open cluster near Delta Cancri. Visible to naked eye.
M97 Ursae Majoris	11	12.6	+55	13	Owl Nebula, diameter 3′. Planetary.
Kappa Crucis, C94	12	50.7	−60	05	'Jewel Box'; open cluster, with stars of contrasting colours.
M3 Can. Ven.	13	40.6	+28	34	Bright globular.
Omega Centauri, C80	13	23.7	−47	03	Finest of all globulars. Easy with naked eye.
M80 Scorpii	16	14.9	−22	53	Globular, between Antares and Beta Scorpii.

Object	RA		Declina-tion		Remarks
	h	m	°	′	
M4 Scorpii	16	21.5	−26	26	Open cluster close to Antares.
M13 Herculis	16	40	+36	31	Globular. Just visible to naked eye.
M92 Herculis	16	16.1	+43	11	Globular. Between Iota and Eta Herculis.
M6 Scorpii	17	36.8	−32	11	Open cluster; naked eye.
M7 Scorpii	17	50.6	−34	48	Very bright open cluster; naked eye.
M23 Sagittarii	17	54.8	−19	01	Open cluster nearly 50′ in diameter.
H IV 37 Draconis, C6	17	58.6	+66	38	Bright planetary.
M8 Sagittarii	18	01.4	−24	23	Lagoon Nebula. Gaseous. Just visible with naked eye.
NGC 6572 Ophiuchi	18	10.9	+06	50	Bright planetary, between Beta Ophiuchi and Zeta Aquilae.
M17 Sagittarii	18	18.8	−16	12	Omega Nebula. Gaseous. Large and bright.
M11 Scuti	18	49.0	−06	19	Wild Duck. Bright open cluster.
M57 Lyrae	18	52.6	+32	59	Ring Nebula. Brightest of planetaries.
M27 Vulpeculae	19	58.1	+22	37	Dumbbell Nebula, near Gamma Sagittae.
H IV 1 Aquarii, C55	21	02.1	−11	31	Bright planetary, near Nu Aquarii.
M15 Pegasi	21	28.3	+12	01	Bright globular, near Epsilon Pegasi.
M39 Cygni	21	31.0	+48	17	Open cluster between Deneb and Alpha Lacertae. Well seen with low powers.

(M = Messier number; NGC = New General Catalogue number; C = Caldwell number.)

Our Contributors

Dr Paul Murdin is Head of Astronomy at the Particle Physics and Astronomy Research Council (PPARC) and Director of Science at the British National Space Centre. He is, of course, one of our most regular contributors.

Lembit Öpik, the grandson of the great astronomer Ernst Julius Öpik, is the Liberal-Democrat Member of Parliament for Montgomeryshire. He was instrumental in persuading the UK Government to set up a Task Force to assess the threat posed by Near Earth Objects.

Dr Colin Taylor is first and foremost an astronomer, but has a lifelong interest in the Plains Indians of North America, and he is a recognized authority on the subject.

David Weldrake graduated with a first-class honours degree in Astrophysics from the University of Hertfordshire in June 2000. He is currently studying for a PhD at the Research School of Astronomy & Astrophysics, Mount Stromlo Observatory in Canberra, Australia.

Professor George Cole is Emeritus Professor of Theoretical Physics in the Faculty of Science at the University of Hull. He is the author of several books on subjects ranging from mathematical techniques to the physics of planetary interiors, and he has published numerous papers in professional journals.

Professor Chris Kitchin was formerly Director of the University of Hertfordshire Observatory. He is an astrophysicist with a great eagerness in encouraging a popular interest in astronomy. He is the author of several books, and appears regularly on television.

Roger O'Brien has a keen interest in astronomy. Graduating in Astronomy with Computing from the University of Hertfordshire in 1995 has enabled him to lecture about various aspects of astronomy to anyone who will listen.

Iain Nicolson is another of our most regular contributors. He was for many years Senior Lecturer in Astronomy at the University of Hertfordshire, but now devotes his time to writing and broadcasting – from his island home off Oban in Scotland.

Dr Fred Watson is Astronomer-in-Charge at the Anglo-Australian Observatory near Coonabarabran, New South Wales, where he is responsible for the scientific output of the Anglo-Australian Telescope and the UK Schmidt Telescope. He also holds the honorary appointment of Associate Professor of Astronomy in the University of Southern Queensland.

Astronomical Societies in the British Isles

British Astronomical Association
Assistant Secretary: Burlington House, Piccadilly, London W1V 9AG.
Meetings: Lecture Hall of Scientific Societies, Civil Service Commission Building, 23 Savile Row, London W1. Last Wednesday each month (Oct.–June), 5 p.m. and some Saturday afternoons.

Association for Astronomy Education
Secretary: Teresa Grafton, The Association for Astronomy Education, c/o The Royal Astronomical Society, Burlington House, Piccadilly, London W1V 0NL.

Astronomy Ireland
Secretary: Tony Ryan, PO Box 2888, Dublin 1, Ireland.
Web site: www.astronomy.ie; *Email:* info@astronomy.ie
Meetings: 2nd Monday of each month. Telescope meetings every clear Saturday.

Federation of Astronomical Societies
Secretary: Clive Down, 10 Glan-y-Llyn, North Cornelly, Bridgend County Borough, CF33 4EF.
Email: clivedown@btinternet.com

Junior Astronomical Society of Ireland
Secretary: K. Nolan, 5 St Patrick's Crescent, Rathcoole, Co. Dublin.
Meetings: The Royal Dublin Society, Ballsbridge, Dublin 4. Monthly.

Society for Popular Astronomy
Secretary: Guy Fennimore, 36 Fairway, Keyworth, Nottingham, NG12 5DU.
Web site: www.popastro.com; *Email:* SPAstronomy@aol.com
Meetings: Last Saturday in Jan., Apr., July, Oct., 2.30 p.m. in London.

Webb Society
Secretary: M. B. Swan, Carrowreagh, Kilshanny, Kilfenora, Co. Clare, Ireland.

Aberdeen and District Astronomical Society
Secretary: Ian C. Giddings, 95 Brentfield Circle, Ellon, Aberdeenshire AB41 9DB.
Meetings: Robert Gordon's Institute of Technology, St Andrew's Street, Aberdeen. Fridays, 7.30 p.m.

Abingdon Astronomical Society (was **Fitzharry's Astronomical Society**)
Secretary: Chris Holt, 9 Rutherford Close, Abingdon, Oxon OX14 2AT.
Web site: www.abingdonastro.org.uk; *Email:* info@abingdonastro.co.uk
Meetings: All Saints' Methodist Church Hall, Dorchester Crescent, Abingdon, Oxon. 2nd Monday Sept.–June, 8 p.m. and additional beginners' meetings and observing evenings as advertised.

Altrincham and District Astronomical Society
Secretary: Derek McComiskey, 33 Tottenham Drive, Manchester M23 9WH.
Meetings: Timperley Village Club. 1st Friday Sept.–June, 8 p.m.

Andover Astronomical Society
Secretary: Mrs S. Fisher, Staddlestones, Aughton, Kingston, Marlborough, Wiltshire, SN8 3SA.
Meetings: Grately Village Hall. 3rd Thursday each month, 7.30 p.m.

Astra Astronomy Section
Secretary: c/o Duncan Lunan, Flat 65, Dalraida House, 56 Blythswood Court, Anderston, Glasgow G2 7PE.
Meetings: Airdrie Arts Centre, Anderson Street, Airdrie. Weekly.

Astrodome Mobile School Planetarium
Contact: Peter J. Golding, 53 City Way, Rochester, Kent ME1 2AX.
Web site: www.astrodome.clara.co.uk; *Email:* astrodome@clara.co.uk

Aylesbury Astronomical Society
Secretary: Alan Smith, 182 Marley Fields, Leighton Buzzard, Beds, LU7 8WN.
Meetings: 1st Monday in month at 8 p.m., venue in Aylesbury area. Details from Secretary.

Bassetlaw Astronomical Society
Secretary: Andrew Patton, 58 Holding, Worksop, Notts S81 0TD.
Meetings: Rhodesia Village Hall, Rhodesia, Worksop, Notts. 2nd and 4th Tuesdays of month at 7.45 p.m.

Batley & Spenborough Astronomical Society
Secretary: Robert Morton, 22 Links Avenue, Cleckheaton, West Yorks BD19 4EG.
Meetings: Milner K. Ford Observatory, Wilton Park, Batley. Every Thursday, 8 p.m.

Bedford Astronomical Society
Secretary: Mrs L. Harrington, 24 Swallowfield, Wyboston, Bedfordshire, MK44 3AE.
Web site: www.observer1.freeserve.co.uk/bashome.html
Meetings: Bedford School, Burnaby Rd, Bedford. Last Wednesday each month.

Bingham & Brooks Space Organization
Secretary: N. Bingham, 15 Hickmore's Lane, Lindfield, W. Sussex.

Birmingham Astronomical Society
Contact: P. Bolas, 4 Moat Bank, Bretby, Burton on Trent DE15 0QJ.
Web site: www.birmingham-astronomical.co.uk; *Email:* pbolas@aol.com
Meetings: Room 146, Aston University. Last Tuesday of month. Sept.–June (except Dec., moved to 1st week in Jan.).

Blackburn Leisure Astronomy Section
Secretary: Mr H. Murphy, 20 Princess Way, Beverley, East Yorkshire, HU17 8PD.
Meetings: Blackburn Leisure Welfare. Mondays, 8 p.m.

Blackpool & District Astronomical Society
Secretary: Terry Devon, 30 Victory Road, Blackpool, Lancashire, FY1 3JT.
Acting Secretary: Tony Evanson, 25 Aintree Road, Thornton, Lancashire, FY5 5HW.
Web site: www.geocities.com/bad_astro/index.html; *Email:* bad_astro@yahoo.co.uk
Meetings: St Kentigens Social Centre, Blackpool. 1st Wednesday of the month, 8 p.m.

Bolton Astronomical Society
Secretary: Peter Miskiw, 9 Hedley Street, Bolton, Lancashire, BL1 3LE.
Meetings: Ladybridge Community Centre, Bolton. 1st and 3rd Tuesdays Sept.–May, 7.30 p.m.

Border Astronomy Society
Secretary: David Pettitt, 14 Sharp Grove, Carlisle, Cumbria, CA2 5QR.
Web site: www.members.aol.com/P3pub/page8.html;
Email: davidpettitt@supanet.com
Meetings: The Observatory, Trinity School, Carlisle. Alternate Thursdays, 7.30 p.m., Sept.–May.

Boston Astronomers
Secretary: Mrs Lorraine Money, 18 College Park, Horncastle, Lincolnshire, LN9 6RE.
Meetings: Blackfriars Arts Centre, Boston. 2nd Monday each month, 7.30 p.m.

Bradford Astronomical Society
Contact: Mrs J. Hilary Knaggs, 6 Meadow View, Wyke, Bradford, BD12 9LA.
Web site: www.bradford-astro.freeserve.co.uk/index.htm
Meetings: Eccleshill Library, Bradford. Alternate Mondays, 7.30 p.m.

Braintree, Halstead & District Astronomical Society
Secretary: Mr J. R. Green, 70 Dorothy Sayers Drive, Witham, Essex, CM8 2LU.
Meetings: BT Social Club Hall, Witham Telephone Exchange. 3rd Thursday each month, 8 p.m.

Breckland Astronomical Society (was **Great Ellingham and District Astronomy Club**)
Contact: Martin Wolton, Willowbeck House, Pulham St Mary, Norfolk, IP21 4QS.
Meetings: Great Ellingham Recreation Centre, Watton Road (B1077), Great Ellingham, 2nd Friday each month, 7.15 p.m.

Bridgend Astronomical Society
Secretary: Clive Down, 10 Glan-y-Llyn, Broadlands, North Cornelly, Bridgend County, CF33 4EF.
Email: clivedown@btinternet.com
Meetings: Bridgend Bowls Centre, Bridgend. 2nd Friday, monthly, 7.30 p.m.

Bridgwater Astronomical Society
Secretary: Mr G. MacKenzie, Watergore Cottage, Watergore, South Petherton, Somerset, TA13 5JQ.
Web site: www.ourworld.compuserve.com/hompages/dbown/Bwastro.htm
Meetings: Room D10, Bridgwater College, Bath Road Centre, Bridgwater. 2nd Wednesday each month, Sept.–June.

Bridport Astronomical Society
Secretary: Mr G. J. Lodder, 3 The Green, Walditch, Bridport, Dorset, DT6 4LB.
Meetings: Walditch Village Hall, Bridport. 1st Sunday each month, 7.30 p.m.

Brighton Astronomical and Scientific Society
Secretary: Ms T. Fearn, 38 Woodlands Close, Peacehaven, East Sussex, BN10 7SF.
Meetings: St Johns Church Hall, Hove. 1st Tuesday each month, 7.30 p.m.

Bristol Astronomical Society
Secretary: Dr John Pickard, 'Fielding', Easter Compton, Bristol, BS35 5SJ.
Meetings: Frank Lecture Theatre, University of Bristol Physics Dept., alternate Fridays in term time, and Westbury Park Methodist Church Rooms, North View, other Fridays.

Cambridge Astronomical Society
Secretary: Brian Lister, 80 Ramsden Square, Cambridge CB4 2BL.
Meetings: Institute of Astronomy, Madingley Road. 3rd Friday each month.

Cardiff Astronomical Society

Secretary: D. W. S. Powell, 1 Tal-y-Bont Road, Ely, Cardiff CF5 5EU.

Meetings: Dept. of Physics and Astronomy, University of Wales, Newport Road, Cardiff. Alternate Thursdays, 8 p.m.

Castle Point Astronomy Club

Secretary: Andrew Turner, 3 Canewdon Hall Close, Canewdon, Rochford, Essex SS4 3PY.

Meetings: St Michael's Church Hall, Daws Heath. Wednesdays, 8 p.m.

Chelmsford Astronomers

Secretary: Brendan Clark, 5 Borda Close, Chelmsford, Essex.

Meetings: Once a month.

Chester Astronomical Society

Secretary: Mrs S. Brooks, 39 Halton Road, Great Sutton, South Wirral, LL66 2UF.

Meetings: All Saints Parish Church, Chester. Last Wednesday each month except Aug. and Dec., 7.30 p.m.

Chester Society of Natural Science, Literature and Art

Secretary: Paul Braid, 'White Wing', 38 Bryn Avenue, Old Colwyn, Colwyn Bay LL29 8AH.

Email: p.braid@virgin.net

Meetings: Once a month.

Chesterfield Astronomical Society

Secretary: Mr Robert McGregor, 34 Higher Albert Street, Chesterfield, Derbyshire, S41 7QE.

Meetings: Barnet Observatory, Newbold, each Friday.

Clacton & District Astronomical Society

Secretary: C. L. Haskell, 105 London Road, Clacton-on-Sea, Essex.

Cleethorpes & District Astronomical Society

Secretary: C. Illingworth, 38 Shaw Drive, Grimsby, S. Humberside.

Meetings: Beacon Hill Observatory, Cleethorpes. 1st Wednesday each month.

Cleveland & Darlington Astronomical Society

Secretary: Neil Haggath, 5 Fountains Crescent, Eston, Middlesborough, Cleveland.

Meetings: Thorpe Thewles Parish Hall, near Stockton on Tees. 2nd Friday, monthly.

Cork Astronomy Club

Secretary: Charles Coughlan, 12 Forest Ridge Crescent, Wilton, Cork, Ireland.

Meetings: 1st Monday, Sept.–May (except bank holidays).

Cornwall Astronomical Society

Secretary: J. M. Harvey, 1 Tregunna Close, Porthleven, Cornwall TR13 9LW.

Meetings: Godolphin Club, Wendron Street, Helston, Cornwall. 2nd and 4th Thursday of each month, 7.30 for 8 p.m.

Cotswold Astronomical Society

Secretary: Rod Salisbury, Grove House, Christchurch Road, Cheltenham, Glos GL50 2PN.

Web site: www.members.nbci.com/CotswoldAS

Meetings: Shurdington Church Hall, School Lane, Shurdington, Cheltenham. 2nd Saturday each month, 8 p.m.

Coventry & Warwickshire Astronomical Society
Secretary: Steve Payne, 68 Stonebury Avenue, Eastern Green, Coventry CV5 7FW.
Web site: www.cawas.freeserve.co.uk; *Email:* sjp2000@thefarside57.freeserve.co.uk
Meetings: The Earlsdon Church Hall, Albany Road, Earlsdon, Coventry. 2nd Friday, monthly, Sept.–June.

Crawley Astronomical Society
Secretary: Ron Gamer, 1 Pevensey Close, Pound Hill, Crawley, West Sussex RH10 7BL.
Meetings: Ifield Community Centre, Ifield Road, Crawley. 3rd Friday each month, 7.30 p.m.

Crayford Manor House Astronomical Society
Secretary: Roger Pickard, 28 Appletons, Hadlow, Kent TM1 0DT.
Meetings: Manor House Centre, Crayford. Monthly during term time.

Croydon Astronomical Society
Secretary: John Murrell, 17 Dalmeny Road, Carshalton, Surrey.
Meetings: Lecture Theatre, Royal Russell School, Combe Lane, South Croydon. Alternate Fridays, 7.45 p.m.

Derby & District Astronomical Society
Secretary: Ian Bennett, Freers Cottage, Sutton Lane, Etwall.
Web site: www.derby-astro-soc.fsnet/index.html;
Email: bennett.lovatt@btinternet.com
Meetings: Friends Meeting House, Derby. 1st Friday each month, 7.30 p.m.

Doncaster Astronomical Society
Secretary: Phil Snowden, 18 Newbury Way, Cusworth, Doncaster DN5.
Meetings: Corporation Brewery Tap public house, 2nd and 4th Wednesdays each month.

Dumfries Astronomical Society
Secretary: Mr J. Sweeney, 3 Lakeview, Powfoot, Annan, DG13 5PG.
Meetings: Gracefield Arts Centre, Edinburgh Road, Dumfries. 3rd Tuesday Aug.–May, 7.30 p.m.

Dundee Astronomical Society
Secretary: G. Young, 37 Polepark Road, Dundee, Tayside, DD1 5QT.
Meetings: Mills Observatory, Balgay Park, Dundee. 1st Friday each month, 7.30 p.m. Sept.–Apr.

Easington and District Astronomical Society
Secretary: T. Bradley, 52 Jameson Road, Hartlepool, Co. Durham.
Meetings: Easington Comprehensive School, Easington Colliery. Every 3rd Thursday throughout the year, 7.30 p.m.

Eastbourne Astronomical Society
Secretary: Peter Gill, 18 Selwyn House, Selwyn Road, Eastbourne, East Sussex BN21 2LF.
Meetings: Willingdon Memorial Hall, Church Street, Willingdon. One Saturday per month, Sept.–July, 7.30 p.m.

East Riding Astronomers
Secretary: Tony Scaife, 15 Beech Road, Elloughton, Brough, North Humberside, HU15 1JX.
Meetings: As arranged.

East Sussex Astronomical Society

Secretary: Marcus Croft, 12 St Marys Cottages, Ninfield Road, Bexhill on Sea, East Sussex.

Web site: www.esas.org.uk

Meetings: St Marys School, Wrestwood Road, Bexhill. 1st Thursday of each month, 8 p.m.

Astronomical Society of Edinburgh

Secretary: Graham Rule, 105/19 Causewayside, Edinburgh EH9 1QG.

Web site: www.roe.ac.uk/asewww/; *Email:* asewww@roe.ac.uk

Meetings: City Observatory, Calton Hill, Edinburgh. 1st Friday each month, 8 p.m.

Edinburgh University Astronomical Society

Secretary: c/o Dept. of Astronomy, Royal Observatory, Blackford Hill, Edinburgh.

Ewell Astronomical Society

Secretary: G. O'Mara, 46 Stanton Close, Epsom KT19 9NP.

Web site: www.ewell-as.co.uk

Meetings: St Mary's Church Hall, London Road, Ewell. 2nd Friday of each month except August, 7.45 p.m.

Exeter Astronomical Society

Secretary: Tim Sedgwick, Old Dower House, Half Moon, Newton St Cyres, Exeter, Devon, EX5 5AE.

Meetings: The Meeting Room, Wynards, Magdalen Street, Exeter. 1st Thursday of month.

Farnham Astronomical Society

Secretary: Laurence Anslow, 'Asterion', 18 Wellington Lane, Farnham, Surrey, GU9 9BA.

Meetings: Central Club, South Street, Farnham. 2nd Thursday each month, 8 p.m.

Foredown Tower Astronomy Group

Secretary: M. Feist, Foredown Tower Camera Obscura, Foredown Road, Portslade, East Sussex BN41 2EW.

Meetings: At the above address, 3rd Tuesday each month. 7 p.m. (winter), 8 p.m. (summer).

Fylde Astronomical Society

Secretary: 28 Belvedere Road, Thornton, Lancs.

Meetings: Stanley Hall, Rossendale Avenue South. 1st Wednesday each month.

Astronomical Society of Glasgow

Secretary: Mr Robert Hughes, Apartment 8/4, 75 Plean Street, Glasgow G14 0YW.

Meetings: University of Strathclyde, George St, Glasgow. 3rd Thursday each month, Sept.–Apr., 7.30 p.m.

Greenock Astronomical Society

Secretary: Carl Hempsey, 49 Brisbane Street, Greenock.

Meetings: Greenock Arts Guild, 3 Campbell Street, Greenock.

Grimsby Astronomical Society

Secretary: R. Williams, 14 Richmond Close, Grimsby, South Humberside.

Meetings: Secretary's home. 2nd Thursday each month, 7.30 p.m.

Guernsey: La Société Guernesiasie Astronomy Section

Secretary: Debby Quertier, Lamorna, Route Charles, St Peter Port, Guernsey GY1 1QS and Jessica Harris, Keanda, Les Sauvagees, St Sampsons, Guernsey GY2 4XT.

Meetings: Observatory, Rue du Lorier, St Peters. Tuesdays, 8 p.m.

Guildford Astronomical Society

Secretary: A. Langmaid, 22 West Mount, The Mount, Guildford, Surrey, GU2 5HL.

Meetings: Guildford Institute, Ward Street, Guildford. 1st Thursday each month, except Aug., 7.30 p.m.

Gwynedd Astronomical Society

Secretary: Mr Ernie Greenwood, 18 Twrcelyn Street, Llanerchymedd, Anglesey LL74 8TL.

Meetings: Dept. of Electronic Engineering, Bangor University. 1st Thursday each month except Aug., 7.30 p.m.

The Hampshire Astronomical Group

Secretary: Geoff Mann, 10 Marie Court, 348 London Road, Waterlooville, Hants PO7 7SR.

Web site: www.hantsastro.demon.co.uk; *Email:* Geoff.Mann@hazleton97.fsnet.co.uk

Meetings: 2nd Friday, Clanfield Memorial Hall, all other Fridays Clanfield Observatory.

Hanney & District Astronomical Society

Secretary: Bob Church, 47 Upthorpe Drive, Wantage, Oxfordshire, OX12 7DG.

Meetings: Last Thursday each month, 8 p.m.

Astronomical Society of Haringey

Secretary: Jerry Workman, 91 Greenslade Road, Barking, Essex, IG11 9XF.

Meetings: Palm Court, Alexandra Palace, 3rd Wednesday each month, 8 p.m.

Harrogate Astronomical Society

Secretary: Brian Bonser, 114 Main Street, Little Ouseburn, TO5 9TG.

Meetings: National Power HQ, Beckwith Knowle, Harrogate. Last Friday each month.

Hastings and Battle Astronomical Society

Secretary: K. A. Woodcock, 24 Emmanuel Road, Hastings, East Sussex, TN34 3LB.

Email: keith@habas.freeserve.co.uk

Meetings: Herstmonceux Science Centre. 2nd Saturday of each month, 7.30 p.m.

Havering Astronomical Society

Secretary: Frances Ridgley, 133 Severn Drive, Upminster, Essex, RM14 1PP.

Meetings: Cranham Community Centre, Marlborough Gardens, Upminster, Essex. 3rd Wednesday each month (except July and Aug.), 7.30 p.m.

Heart of England Astronomical Society

Secretary: John Williams, 100 Stanway Road, Shirley, Solihull, B90 3JG.

Web site: www.members.aol.com/hoeas/home.html; *Email:* hoeas@aol.com

Meetings: Furnace End Village, over Whitacre, Warwickshire. Last Thursday each month, except June, July & Aug., 8 p.m.

Hebden Bridge Literary & Scientific Society, Astronomical Section

Secretary: Peter Jackson, 44 Gilstead Lane, Bingley, West Yorkshire, BD16 3NP.

Meetings: Hebden Bridge Information Centre. Last Wednesday, Sept.–May.

Herschel Astronomy Society

Secretary: Kevin Bishop, 106 Holmsdale, Crown Wood, Bracknell, Berkshire, RG12 3TB.

Meetings: Eton College. 2nd Friday each month, 7.30 p.m.

Highlands Astronomical Society

Secretary: Richard Green, 11 Drumossie Avenue, Culcabock, Inverness IV2 3SJ.

Meetings: The Spectrum Centre, Inverness. 1st Tuesday each month, 7.30 p.m.

Hinckley & District Astronomical Society

Secretary: Mr S. Albrighton, 4 Walnut Close, The Bridleways, Hartshill, Nuneaton, Warwickshire, CV10 0XH.

Meetings: Burbage Common Visitors Centre, Hinckley. 1st Tuesday Sept.–May, 7.30 p.m.

Horsham Astronomy Group (was **Forest Astronomical Society**)

Secretary: Mr A. R. Clarke, 93 Clarence Road, Horsham, West Sussex, RH13 5SL.

Meetings: 1st Wednesday each month.

Howards Astronomy Club

Secretary: H. Ilett, 22 St Georges Avenue, Warblington, Havant, Hants.

Meetings: To be notified.

Huddersfield Astronomical and Philosophical Society

Secretary: Lisa B. Jeffries, 58 Beaumont Street, Netherton, Huddersfield, West Yorkshire, HD4 7HE.

Email: l.b.jeffries@hud.ac.uk

Meetings: 4a Railway Street, Huddersfield. Every Wednesday and Friday, 7.30 p.m.

Hull and East Riding Astronomical Society

Secretary: Tony Scaife, 15 Beech Road, Elloughton, Brough, North Humberside, HU15 1JX.

Meetings: Wyke 6th Form College, Bricknell Avenue, Hull. 2nd Tuesday each month, Oct.–Apr., 7.30 p.m.

Ilkeston & District Astronomical Society

Secretary: Mark Thomas, 2 Elm Avenue, Sandiacre, Nottingham NG10 5EJ.

Meetings: The Function Room, Erewash Museum, Anchor Row, Ilkeston. 2nd Tuesday monthly, 7.30 p.m.

Ipswich, Orwell Astronomical Society

Secretary: R. Gooding, 168 Ashcroft Road, Ipswich.

Meetings: Orwell Park Observatory, Nacton, Ipswich. Wednesdays, 8 p.m.

Irish Astronomical Association

Secretary: Terry Moseley (President), 6 Collinbridge Drive, Newtownabbey, Co. Antrim BT36 7SX.

Email: terrymosel@aol.com

Meetings: Ashby Building, Stranmillis Road, Belfast. Alternate Wednesdays, 7.30 p.m.

Irish Astronomical Society

Secretary: James O'Connor, PO Box 2547, Dublin 15, Ireland.

Meetings: Ely House, 8 Ely Place, Dublin 2. 1st and 3rd Monday each month.

Isle of Man Astronomical Society

Secretary: James Martin, Ballaterson Farm, Peel, Isle of Man IM5 3AB.

Email: ballaterson@manx.net

Meetings: Isle of Man Observatory, Foxdale. 1st Thursday of each month, 8 p.m.

Isle of Wight Astronomical Society

Secretary: J. W. Feakins, 1 Hilltop Cottages, High Street, Freshwater, Isle of Wight.

Meetings: Unitarian Church Hall, Newport, Isle of Wight. Monthly.

Keele Astronomical Society

Secretary: Natalie Webb, Department of Physics, University of Keele, Keele, Staffordshire, ST5 5BG.

Meetings: As arranged during term time.

Kettering and District Astronomical Society
Asst. *Secretary:* Steve Williams, 120 Brickhill Road, Wellingborough, Northants.
Meetings: Quaker Meeting Hall, Northall Street, Kettering, Northants. 1st Tuesday each month, 7.45 p.m.

King's Lynn Amateur Astronomical Association
Secretary: P. Twynman, 17 Poplar Avenue, RAF Marham, King's Lynn.
Meetings: As arranged.

Lancaster and Morecambe Astronomical Society
Secretary: Mrs E. Robinson, 4 Bedford Place, Lancaster, LA1 4EB.
Email: ehelenerob@btinternet.com
Meetings: Church of the Ascension, Torrisholme. 1st Wednesday each month, except July and Aug.

Lancaster University Astronomical Society
Secretary: c/o Students Union, Alexandra Square, University of Lancaster.
Meetings: As arranged.

Laymans Astronomical Society
Secretary: John Evans, 10 Arkwright Walk, The Meadows, Nottingham.
Meetings: The Popular, Bath Street, Ilkeston, Derbyshire. Monthly.

Leeds Astronomical Society
Secretary: Mark A. Simpson, 37 Roper Avenue, Gledhow, Leeds, LS8 1LG.
Meetings: Centenary House, North Street. 2nd Wednesday each month, 7.30 p.m.

Leicester Astronomical Society
Secretary: Dr P. J. Scott, 21 Rembridge Close, Leicester LE3 9AP.
Meetings: Judgemeadow Community College, Marydene Drive, Evington, Leicester. 2nd and 4th Tuesdays each month, 7.30 p.m.

Letchworth and District Astronomical Society
Secretary: Eric Hutton, 14 Folly Close, Hitchin, Herts.
Meetings: As arranged.

Lewes Amateur Astronomers
Secretary: Christa Sutton, 8 Tower Road, Lancing, West Sussex, BN15 9HT.
Meetings: The Bakehouse Studio, Lewes. Last Wednesday each month.

Limerick Astronomy Club
Secretary: Tony O'Hanlon, 26 Ballycannon Heights, Meelick, Co. Clare, Ireland.
Meetings: Limerick Senior College, Limerick, Ireland. Monthly (except June and Aug.), 8 p.m.

Lincoln Astronomical Society
Secretary: David Swaey, 'Everglades', 13 Beaufort Close, Lincoln LN2 4SF.
Meetings: The Lecture Hall, off Westcliffe Street, Lincoln. 1st Tuesday each month.

Liverpool Astronomical Society
Secretary: Mr K. Clark, 31 Sandymount Drive, Wallasey, Merseyside L45 0LJ.
Meetings: Lecture Theatre, Liverpool Museum. 3rd Friday each month, 7 p.m.

Norman Lockyer Observatory Society
Secretary: G. E. White, PO Box 9, Sidmouth EX10 0YQ.
Web site: www.ex.ac.uk/nlo/; *Email:* g.e.white@ex.ac.uk
Meetings: Norman Lockyer Observatory, Sidmouth. Fridays and 2nd Monday each month, 7.30 p.m.

Loughton Astronomical Society
Secretary: Charles Munton, 14a Manor Road, Wood Green, London N22 4YJ.
Meetings: 1st Theydon Bois Scout Hall, Loughton Lane, Theydon Bois. Weekly.

Lowestoft and Great Yarmouth Regional Astronomers (LYRA) Society

Secretary: Simon Briggs, 28 Sussex Road, Lowestoft, Suffolk.

Meetings: Community Wing, Kirkley High School, Kirkley Run, Lowestoft. 3rd Thursday each month, 7.30 p.m.

Luton Astronomical Society

Secretary: Mr G. Mitchell, Putteridge Bury, University of Luton, Hitchin Road, Luton.

Web site: www.lutonastrosoc.org.uk; *Email:* user998491@aol.com

Meetings: Putteridge Bury, Luton. Last Friday each month, 7.30 p.m.

Lytham St Annes Astronomical Association

Secretary: K. J. Porter, 141 Blackpool Road, Ansdell, Lytham St Annes, Lancs.

Meetings: College of Further Education, Clifton Drive South, Lytham St Annes. 2nd Wednesday monthly Oct.–June.

Macclesfield Astronomical Society

Secretary: Mr John H. Thomson, 27 Woodbourne Road, Sale, Chesire M33 3SY

Web site: www.g0-evp.demon.co.uk; *Email:* jhandlc@yahoo.com

Meetings: Jodrell Bank Science Centre, Goostrey, Cheshire. 1st Tuesday of every month, 7 p.m.

Maidenhead Astronomical Society

Secretary: Tim Haymes, Hill Rise, Knowl Hill Common, Knowl Hill, Reading RG10 9YD.

Meetings: Stubbings Church Hall, near Maidenhead. 1st Friday Sept.–June.

Maidstone Astronomical Society

Secretary: Stephen James, 4 The Cherry Orchard, Haddow, Tonbridge, Kent.

Meetings: Nettlestead Village Hall. 1st Tuesday in the month except July and Aug., 7.30 p.m.

Manchester Astronomical Society

Secretary: Mr Kevin J. Kilburn FRAS, Godlee Observatory, UMIST, Sackville Street, Manchester M60 1QD.

Web site: www.u-net.com/ph/mas/; *Email:* kkilburn@globalnet.co.uk

Meetings: At the Godlee Observatory. Thursdays, 7 p.m., except below.

Free Public Lectures: Renold Building UMIST, third Thursday Sept.–Mar., 7.30 p.m.

Mansfield and Sutton Astronomical Society

Secretary: Angus Wright, Sherwood Observatory, Coxmoor Road, Sutton-in-Ashfield, Nottinghamshire NG17 5LF.

Meetings: Sherwood Observatory, Coxmoor Road. Last Tuesday each month, 7.30 p.m.

Mexborough and Swinton Astronomical Society

Secretary: Mark R. Benton, 14 Sandalwood Rise, Swinton, Mexborough, South Yorkshire, S64 8PN.

Web site: www.msas.org.uk; *Email:* mark@masas.f9.co.uk

Meetings: Swinton WMC. Thursdays, 7.30 p.m.

Mid-Kent Astronomical Society

Secretary: Peter Bassett, 167 Shakespeare Road, Gillingham, Kent, ME7 5QB.

Meetings: Riverside Country Park, Lower Rainham Road, Gillingham. 2nd and last Fridays each month, 7.45 p.m.

Milton Keynes Astronomical Society

Secretary: Mike Leggett, 19 Matilda Gardens, Shenley Church End, Milton Keynes, MK5 6HT.

Web site: www.mkas.org.uk; *Email:* mike-pat-leggett@shenley9.fsnet.co.uk

Meetings: Rectory Cottage, Bletchley. Alternate Fridays.

Moray Astronomical Society

Secretary: Richard Pearce, 1 Forsyth Street, Hopeman, Elgin, Moray, Scotland.

Meetings: Village Hall Close, Co. Elgin.

Newbury Amateur Astronomical Society

Secretary: Miss Nicola Evans, 'Romaron', Bunces Lane, Burghfield Common, Reading RG7 3DG.

Meetings: United Reformed Church Hall, Cromwell Place, Newbury. 2nd Friday of month, Sept.–June.

Newcastle-on-Tyne Astronomical Society

Secretary: C. E. Willits, 24 Acomb Avenue, Seaton Delaval, Tyne and Wear.

Meetings: Zoology Lecture Theatre, Newcastle University. Monthly.

North Aston Space & Astronomical Club

Secretary: W. R. Chadburn, 14 Oakdale Road, North Aston, Sheffield.

Meetings: To be notified.

Northamptonshire Natural History Society (Astronomy Section)

Secretary: R. A. Marriott, 24 Thirlestane Road, Northampton NN4 8HD.

Email: ram@hamal.demon.co.uk

Meetings: Humfrey Rooms, Castilian Terrace, Northampton. 2nd and last Mondays, most months, 7.30 p.m.

Northants Amateur Astronomers

Secretary: Mervyn Lloyd, 76 Havelock Street, Kettering, Northamptonshire.

Meetings: 1st and 3rd Tuesdays each month, 7.30 p.m.

North Devon Astronomical Society

Secretary: P. G. Vickery, 12 Broad Park Crescent, Ilfracombe, Devon, EX34 8DX.

Meetings: Methodist Hall, Rhododendron Avenue, Sticklepath, Barnstaple. 1st Wednesday each month, 7.15 p.m.

North Dorset Astronomical Society

Secretary: J. E. M. Coward, The Pharmacy, Stalbridge, Dorset.

Meetings: Charterhay, Stourton, Caundle, Dorset. 2nd Wednesday each month.

North Downs Astronomical Society

Secretary: Martin Akers, 36 Timber Tops, Lordswood, Chatham, Kent, ME5 8XQ.

Meetings: Vigo Village Hall. 3rd Thursday each month. 7.30 p.m.

North-East London Astronomical Society

Secretary: Mr B. Beeston, 38 Abbey Road, Bush Hill Park, Enfield EN1 2QN.

Meetings: Wanstead House, The Green, Wanstead. 3rd Sunday each month (except Aug.), 3 p.m.

North Gwent and District Astronomical Society

Secretary: Jonathan Powell, 14 Lancaster Drive, Gilwern, nr Abergavenny, Monmouthshire, NP7 0AA.

Meetings: Gilwern Community Centre. 15th of each month, 7.30 p.m.

North Staffordshire Astronomical Society

Secretary: Duncan Richardson, Halmerend Hall Farm, Halmerend, Stoke-on-Trent, Staffordshire, ST7 8AW.

Email: dwr@enterprise.net

Meetings: 21st Hartstill Scout Group HQ, Mount Pleasant, Newcastle-under-Lyme ST5 1DR. 1st Tuesday each month (except July and Aug.), 7–9.30 p.m.

North Western Association of Variable Star Observers

Secretary: Jeremy Bullivant, 2 Beaminster Road, Heaton Mersey, Stockport, Cheshire.

Meetings: Four annually.

Norwich Astronomical Society

Secretary: Frank Lawlor, 'Farnworth', Poringland Road, Upper Stoke Holy Cross, Norwich NR14 8NW.

Web site: www.nas.gurney.org.uk

Meetings: Seething Observatory, Toad Lane, Thwaite St Mary, Norfolk. Every Friday, 7.30 p.m.

Nottingham Astronomical Society

Secretary: C. Brennan, 40 Swindon Close, The Vale, Giltbrook, Nottingham NG16 2WD.

Meetings: Djanogly City Technology College, Sherwood Rise (B682). 1st and 3rd Thursdays each month, 7.30 p.m.

Oldham Astronomical Society

Secretary: P. J. Collins, 25 Park Crescent, Chadderton, Oldham.

Meetings: Werneth Park Study Centre, Frederick Street, Oldham. Fortnightly, Friday.

Open University Astronomical Society

Secretary: Dr Andrew Norton, Department of Physics and Astronomy, The Open University, Walton Hall, Milton Keynes MK7 6AA.

Web site: www.physics.open.ac.uk/research/astro/a_club.html

Meetings: Open University, Milton Keynes. 1st Tuesday of every month, 7.30 p.m.

Orpington Astronomical Society

Secretary: Dr Ian Carstairs, 38 Brabourne Rise, Beckenham, Kent BR3 2SG.

Meetings: High Elms Nature Centre, High Elms Country Park, High Elms Road, Farnborough, Kent. 4th Thursday each month, Sept.–July, 7.30 p.m.

Papworth Astronomy Club

Contact: Keith Tritton, Magpie Cottage, Fox Street, Great Gransden, Sandy, Bedfordshire SG19 3AA.

Email: kpt2@tutor.open.ac.uk

Meetings: Bradbury Progression Centre, Church Lane, Papworth Everard, near Huntingdon. 1st Wednesday each month, 7 p.m.

Peterborough Astronomical Society

Secretary: Sheila Thorpe, 6 Cypress Close, Longthorpe, Peterborough.

Meetings: 1st Thursday every month, 7.30 p.m.

Plymouth Astronomical Society

Secretary: Alan G. Penman, 12 St Maurice View, Plympton, Plymouth, Devon PL7 1FQ.

Email: oakmount12@aol.com

Meetings: Glynis Kingham Centre, YMCA Annex, Lockyer Street, Plymouth. 2nd Friday each month, 7.30 p.m.

PONLAF
Secretary: Matthew Hepburn, 6 Court Road, Caterham, Surrey CR3 5RD.
Meetings: Room 5, 6th floor, Tower Block, University of North London. Last Friday each month during term time, 6.30 p.m.

Port Talbot Astronomical Society (was **Astronomical Society of Wales**)
Secretary: Mr J. Hawes, 15 Lodge Drive, Baglan, Port Talbot, West Glamorgan SA12 8UD.
Meetings: Port Talbot Arts Centre. 1st Tuesday each month, 7.15 p.m.

Portsmouth Astronomical Society
Secretary: G. B. Bryant, 81 Ringwood Road, Southsea.
Meetings: Monday, fortnightly.

Preston & District Astronomical Society
Secretary: P. Sloane, 77 Ribby Road, Wrea Green, Kirkham, Preston, Lancs.
Meetings: Moor Park (Jeremiah Horrocks) Observatory, Preston. 2nd Wednesday, last Friday each month, 7.30 p.m.

Reading Astronomical Society
Secretary: Mrs Ruth Sumner, 22 Anson Crescent, Shinfield, Reading RG2 8JT.
Meetings: St Peter's Church Hall, Church Road, Earley. 3rd Friday each month, 7 p.m.

Renfrewshire Astronomical Society
Secretary: Ian Martin, 10 Aitken Road, Hamilton, South Lanarkshire ML3 7YA.
Web site: www.renfrewshire-as.co.uk; *Email:* RenfrewAS@aol.com
Meetings: Coats Observatory, Oakshaw Street, Paisley. Fridays, 7.30 p.m.

Richmond & Kew Astronomical Society
Secretary: Stewart McLaughlin, 41a Bruce Road, Mitcham, Surrey CR4 2BJ.
Meetings: Richmond Adult College, Parkshot, Richmond, Surrey, and the King's Observatory, Old Deer Park, Richmond, Surrey. Bimonthly.

Rower Astronomical Society
Secretary: Mary Kelly, Knockatore, The Rower, Thomastown, Co. Kilkenny, Ireland.

St Helens Amateur Astronomical Society
Secretary: Carl Dingsdale, 125 Canberra Avenue, Thatto Heath, St Helens, Merseyside WA9 5RT.
Meetings: As arranged.

Salford Astronomical Society
Secretary: Mrs Kath Redford, 2 Albermarle Road, Swinton, Manchester M27 5ST.
Meetings: The Observatory, Chaseley Road, Salford. Wednesdays.

Salisbury Astronomical Society
Secretary: Mrs R. Collins, 3 Fairview Road, Salisbury, Wiltshire, SP1 1JX.
Meetings: Glebe Hall, Winterbourne Earls, Salisbury. 1st Tuesday each month.

Sandbach Astronomical Society
Secretary: Phil Benson, 8 Gawsworth Drive, Sandbach, Cheshire.
Meetings: Sandbach School, as arranged.

Sawtry & District Astronomical Society
Secretary: Brooke Norton, 2 Newton Road, Sawtry, Huntingdon, Cambridgeshire, PE17 5UT.
Meetings: Greenfields Cricket Pavilion, Sawtry Fen. Last Friday each month.

Scarborough & District Astronomical Society
Secretary: Mrs S. Anderson, Basin House Farm, Sawdon, Scarborough, N. Yorks.
Meetings: Scarborough Public Library. Last Saturday each month, 7–9 p.m.

Scottish Astronomers Group
Secretary: Dr Ken Mackay, Hayford House, Cambusbarron, Stirling, FK7 9PR.
Meetings: North of Hadrian's Wall, twice yearly.

Sheffield Astronomical Society
Secretary: Mr Andrew Green, 11 Lyons Street, Ellesmere, Sheffield S4 7QS.
Web site: www.saqqara.demon.co.uk/sas/sashome.htm
Meetings: Twice monthly at Mayfield Environmental Education Centre, David Lane, Fulwood, Sheffield S10, 7.30–10 p.m.

Shetland Astronomical Society
Secretary: Peter Kelly, The Glebe, Fetlar, Shetland, ZE2 9DJ.
Email: theglebe@zetnet.co.uk
Meetings: Fetlar, Fridays, Oct.–Mar.

Shropshire Astronomical Society
Secretary: Mrs Jacqui Dodds, 35 Marton Drive, Wellington, Telford, TF1 3HL.
Web site: www.astro.cf.ac.uk/sas/sasmain.html; *Email:* jacquidodds@ntlworld.com
Meetings: Gateway Arts and Education Centre, Chester Street, Shrewsbury. Occasional Fridays plus monthly observing meetings, Rodington Village Hall.

Sidmouth and District Astronomical Society
Secretary: M. Grant, Salters Meadow, Sidmouth, Devon.
Meetings: Norman Lockyer Observatory, Salcombe Hill. 1st Monday in each month.

Skipton & Craven Astronomical Society
Contact: Tony Ireland, 14 Cross Bank, Skipton, North Yorkshire BD23 6AH.
Email: sacas@andybat.demon.co.uk
Meetings: 3rd Wednesday of each month, Sept.–May.

Solent Amateur Astronomers
Secretary: Ken Medway, 443 Burgess Road, Swaythling, Southampton SO16 3BL.
Web site: www.delscope.demon.co.uk;
Email: kenmedway@kenmedway.demon.co.uk
Meetings: Room 8, Oaklands, Community School, Fairisle Road, Lordshill, Southampton. 3rd Tuesday each month, 7.30 p.m.

Southampton Astronomical Society
Secretary: John Thompson, 4 Heathfield, Hythe, Southampton, SO45 5BJ.
Web site: www.home.clara.net/lmhobbs/sas.html;
Email: John.G.Thompson@Tesco.net
Meetings: Conference Room 3, The Civic Centre, Southampton. 2nd Thursday each month (except Aug.), 7.30 p.m.

South Downs Astronomical Society
Secretary: J. Green, 46 Central Avenue, Bognor Regis, West Sussex, PO21 5HH.
Web site: www.southdowns.org.uk
Meetings: Assembly Rooms, Chichester. 1st Friday in each month.

South-East Essex Astronomical Society
Secretary: C. P. Jones, 29 Buller Road, Laindon, Essex.
Web site: www.seeas.dabsol.co.uk/; *Email:* cpj@cix.co.uk
Meetings: Lecture Theatre, Central Library, Victoria Avenue, Southend-on-Sea. Generally 1st Thursday in month, Sept.–May, 7.30 p.m.

South-East Kent Astronomical Society
Secretary: Andrew McCarthy, 25 St Paul's Way, Sandgate, near Folkestone, Kent, CT20 3NT.
Meetings: Monthly.

South Lincolnshire Astronomical & Geophysical Society

Secretary: Ian Farley, 12 West Road, Bourne, Lincolnshire, PE10 9PS.

Meetings: Adult Education Study Centre, Pinchbeck. 3rd Wednesday each month, 7.30 p.m.

Southport Astronomical Society

Secretary: Patrick Brannon, Willow Cottage, 90 Jacksmere Lane, Scarisbrick, Ormskirk, Lancashire, L40 9RS.

Meetings: Monthly Sept.–May, plus observing sessions.

Southport, Ormskirk and District Astronomical Society

Secretary: J. T. Harrison, 92 Cottage Lane, Ormskirk, Lancs L39 3NJ.

Meetings: Saturday evenings, monthly as arranged.

South Shields Astronomical Society

Secretary: c/o South Tyneside College, St George's Avenue, South Shields.

Meetings: Marine and Technical College. Each Thursday, 7.30 p.m.

South Somerset Astronomical Society

Secretary: G. McNelly, 11 Laxton Close, Taunton, Somerset.

Meetings: Victoria Inn, Skittle Alley, East Reach, Taunton, Somerset. Last Saturday each month, 7.30 p.m.

South-West Hertfordshire Astronomical Society

Secretary: Tom Walsh, 'Finches', Coleshill Lane, Winchmore Hill, Amersham, Buckinghamshire HP7 0NP.

Meetings: Rickmansworth. Last Friday each month, Sept.–May.

Stafford and District Astronomical Society

Secretary: Miss L. Hodkinson, 6 Elm Walk, Penkridge, Staffordshire, ST19 5NL.

Meetings: Weston Road High School, Stafford. Every 3rd Thursday, Sept.–May, 7.15 p.m.

Stirling Astronomical Society

Secretary: Hamish MacPhee, 10 Causewayhead Road, Stirling FK9 5ER.

Meetings: Smith Museum & Art Gallery, Dumbarton Road, Stirling. 2nd Friday each month, 7.30 p.m.

Stoke-on-Trent Astronomical Society

Secretary: M. Pace, Sundale, Dunnocksfold, Alsager, Stoke-on-Trent.

Meetings: Cartwright House, Broad Street, Hanley. Monthly.

Stratford-upon-Avon Astronomical Society

Secretary: Robin Swinbourne, 18 Old Milverton, Leamington Spa, Warwickshire, CV32 6SA.

Meetings: Tiddington Home Guard Club. 4th Tuesday each month, 7.30 p.m.

Sunderland Astronomical Society

Contact: Don Simpson, 78 Stratford Avenue, Grangetown, Sunderland SR2 8RZ.

Meetings: Friends Meeting House, Roker. 1st, 2nd and 3rd Sundays each month.

Sussex Astronomical Society

Secretary: Mrs C. G. Sutton, 75 Vale Road, Portslade, Sussex.

Meetings: English Language Centre, Third Avenue, Hove. Every Wednesday, 7.30–9.30 p.m., Sept.–May.

Swansea Astronomical Society

Secretary: Maurice Convey, 132 Eaton Crescent, Uplands, Swansea SA1 4QR.

Meetings: Lecture Room C, Mathematics and Physics Building, University of Swansea. 2nd and 4th Thursday each month, 7 p.m.

Tavistock Astronomical Society

Secretary: Mrs Ellie Coombes, Rosemount, Under Road, Gunnislake, Cornwall PL18 9JL.

Meetings: Science Laboratory, Kelly College, Tavistock. 1st Wednesday each month, 7.30 p.m.

Thames Valley Astronomical Group

Secretary: K. J. Pallet, 82a Tennyson Street, South Lambeth, London SW8 3TH.

Meetings: As arranged.

Thanet Amateur Astronomical Society

Secretary: P. F. Jordan, 85 Crescent Road, Ramsgate.

Meetings: Hilderstone House, Broadstairs, Kent. Monthly.

Torbay Astronomical Society

Secretary: Tim Moffat, 31 Netley Road, Newton Abbot, Devon, TQ12 2LL.

Meetings: Torquay Boys' Grammar School, 1st Thursday in month; and Town Hall, Torquay, 3rd Thursday in month, Oct.–May, 7.30 p.m.

Tullamore Astronomical Society

Secretary: Tom Walsh, 25 Harbour Walk, Tullamore, Co. Offaly, Ireland.

Web site: www.iol.ie/seanmck/tas.htm; *Email:* tcwalsh25@yahoo.co.uk

Meetings: Order of Malta Lecture Hall, Tanyard, Tullamore, Co. Offaly, Ireland. Mondays at 8 p.m., every fortnight.

Tyrone Astronomical Society

Secretary: John Ryan, 105 Coolnafranky Park, Cookstown, Co. Tyrone.

Meetings: Contact Secretary.

Usk Astronomical Society

Secretary: Bob Wright, 'Llwyn Celyn', 75 Woodland Road, Croesyceiliog, Cwmbran, NP44 2OX.

Meetings: Usk Community Education Centre, Maryport Street, Usk. Every Thursday during school term, 7 p.m.

Vectis Astronomical Society

Secretary: Rosemary Pears, 1 Rockmount Cottages, Undercliff Drive, St Lawrence, Ventnor, Isle of Wight PO38 1XG.

Web site: www.wightskies.fsnet.co.uk/main.html;

Email: may@tatemma.freeserve.co.uk

Meetings: Lord Louis Library Meeting Room, Newport. 4th Friday each month except Dec., 7.30 p.m.

Vigo Astronomical Society

Secretary: Robert Wilson, 43 Admers Wood, Vigo Village, Meopham, Kent DA13 0SP.

Meetings: Vigo Village Hall. As arranged.

Walsall Astronomical Society

Secretary: Bob Cleverley, 40 Mayfield Road, Sutton Coldfield, B74 3PZ.

Meetings: Freetrade Inn, Wood Lane, Pelsall North Common. Every Thursday.

Wellingborough District Astronomical Society

Secretary: S. M. Williams, 120 Brickhill Road, Wellingborough, Northants.

Meetings: Gloucester Hall, Church Street, Wellingborough. 2nd Wednesday each month, 7.30 p.m.

Wessex Astronomical Society

Secretary: Leslie Fry, 14 Hanhum Road, Corfe Mullen, Dorset.

Meetings: Allendale Centre, Wimborne, Dorset. 1st Tuesday of each month.

West Cornwall Astronomical Society

Secretary: Dr R. Waddling, The Pines, Pennance Road, Falmouth, Cornwall TR11 4ED.

Meetings: Helston Football Club, 3rd Thursday each month, and St Michalls Hotel, 1st Wednesday each month, 7.30 p.m.

West of London Astronomical Society

Secretary: Duncan Radbourne, 28 Tavistock Road, Edgware, Middlesex HA8 6DA.

Web site: www.wocas.org.uk

Meetings: Monthly, alternately in Uxbridge and North Harrow. 2nd Monday in month, except Aug.

West Midlands Astronomical Association

Secretary: Miss S. Bundy, 93 Greenridge Road, Handsworth Wood, Birmingham.

Meetings: Dr Johnson House, Bull Street, Birmingham. As arranged.

West Yorkshire Astronomical Society

Secretary: Pete Lunn, 21 Crawford Drive, Wakefield, West Yorkshire.

Meetings: Rosse Observatory, Carleton Community Centre, Carleton Road, Pontefract. Each Tuesday, 7.15 p.m.

Whitby and District Astronomical Society

Secretary: Rosemary Bowman, The Cottage, Larpool Drive, Whitby, North Yorkshire, YO22 4ND.

Meetings: Whitby Mission, Seafarer's Centre, Haggersgate, Whitby. 1st Tuesday of the month, 7.30 p.m.

Whittington Astronomical Society

Secretary: Peter Williamson, The Observatory, Top Street, Whittington, Shropshire.

Meetings: The Observatory. Every month.

Wiltshire Astronomical Society

Secretary: Simon Barnes, 25 Woodcombe, Melksham, Wilts SN12 6HA.

Meetings: St Andrews Church Hall, Church Lane, off Forest Road, Melksham, Wilts.

Wolverhampton Astronomical Society

Secretary: Mr M. Bryce, Iona, 16 Yellowhammer Court, Kidderminster, Worcestershire, DY10 4RR.

Web site: www.wolvas.org.uk; *Email:* michaelbryce@wolvas.org.uk

Meetings: Beckminster Methodist Church Hall, Birches Barn Road, Wolverhampton. Alternate Mondays, Sept.–Apr., extra dates in summer, 7.30 p.m.

Worcester Astronomical Society

Secretary: Mr S. Bateman, 12 Bozward Street, Worcester WR2 5DE.

Meetings: Room 117, Worcester College of Higher Education, Henwick Grove, Worcester. 2nd Thursday each month, 8 p.m.

Worthing Astronomical Society

Contact: G. Boots, 101 Ardingly Drive, Worthing, West Sussex, BN12 4TW.

Web site: www.worthingastro.freeserve.co.uk;

Email: gboots@observatory99.freeserve.co.uk

Meetings: Heene Church Rooms, Heene Road, Worthing. 1st Wednesday each month (except Aug.), 7.30 p.m.

Wycombe Astronomical Society

Secretary: Mr P. Treherne, 34 Honeysuckle Road, Widmer End, High Wycombe, Buckinghamshire, HP15 6BW.

Meetings: Woodrow High House, Amersham. 3rd Wednesday each month, 7.45 p.m.

The York Astronomical Society

Secretary: Mel Pickard, 19 Elmfield Avenue, York YO31 9LX.

Web site: www.yorkastro.freeserve.co.uk; *Email:* yas@talk21.com

Meetings: The Knavesmire Room, York Priory Street Centre, Priory Street, York. 1st and 3rd Friday of each month (except Aug.), 8 p.m.

Any society wishing to be included in this list of local societies or to update details, including any web-site addresses, is invited to write to the Editor (c/o Pan Macmillan, 20 New Wharf Road, London N1 9RR), so that the relevant information may be included in the next edition of the *Yearbook.*

The William Herschel Society maintains the museum established at 19 New King Street, Bath BA1 2BL – the only surviving Herschel House. It also undertakes activities of various kinds. New members would be welcome; those interested are asked to contact the Membership Secretary at the museum.